Praise for *The Fever Trail*

"Meticulous . . . Honigsbaum offers much to ponder about how the first world should be dealing with a serious disease like malaria. . . . His story bursts with facts."
—*The Village Voice*

"An exciting tale, brimming with the exploits of swashbuckling botanists and touching on quinine's vital importance in colonial warfare."
—*Mother Jones*

"A masterful tale of high adventure and scientific discovery."
—Andrew Spielman, Professor of Tropical Public Health, Harvard School of Public Health, and co-author of *Mosquito*

"*The Fever Trail* is an engaging account of one of the great sagas of plant exploration . . . It is a rich and astonishing tapestry, a story of adventure and intrigue, of wealth and power, played out against a backdrop of pharmacological wizardry and botanical splendor."
—Wade Davis, Explorer-in-Residence, National Geographic Society

"A stunning history of the hunt for a cure for malaria."
—*The Daily Telegraph* (London)

"A travel extravaganza . . . a learned work of social and natural history."
—*The Spectator* (London)

"A fascinating story . . . neatly written, authoritative."
—*The Observer* (London)

The Fever Trail

In Search of the Cure for Malaria

Mark Honigsbaum

Picador

Farrar, Straus and Giroux
New York

 For information, address Picador, 175 Fifth Avenue, New York, N.Y. 10010.

www.picadorusa.com

Picador® is a U.S. registered trademark and is used by Farrar, Straus and Giroux under license from Pan Books Limited.

For information on Picador Reading Group Guides, as well as ordering, please contact the Trade Marketing department at St. Martin's Press.
Phone: 1-800-221-7945 extension 763
Fax: 212-677-7456
E-mail: *trademarketing@stmartins.com*

Library of Congress Cataloging-in-Publication Data

Honigsbaum, Mark.
The fever trail : in search of the cure for malaria / Mark Honigsbaum.
p. cm.
Previously published: London : Macmillan, 2001.
Previously published with subtitle: the hunt for the cure for malaria.
Includes bibliographical references and index.
ISBN 0-312-42180-X
1. Malaria—History. I. Title.

RA644.M2 H665 2002
616.9'362'009—dc21 2001040871

First published in Great Britain by Macmillan

First published in the United States of America by Farrar, Straus and Giroux

First Picador Edition: May 2003

10 9 8 7 6 5 4 3 2 1

For Jeanette

Contents

Preface

To the amazement of all, she recovered sooner than you can say it.

—SEBASTIANO BADO,

Anastasis Corticis Peruviae Seu Chinae Chinae Defensio

FOR THE FIRST FIVE THOUSAND YEARS or so of recorded history, no one knew where malaria came from or how to cure it. All that was known was that the fever was swift, often lethal, and left survivors weak and debilitated for months. Then, in 1663, a fantastical story began circulating in the salons of Europe. It told of a beautiful woman, Doña Francisca Henriquez de Ribera, the fourth Condesa (Countess) de Chinchón, and her miraculous cure from an intermittent fever in Lima, then part of the Spanish Viceroyalty of Peru.

In 1638, went the story, the condesa contracted a particularly virulent strain of the parasitic illness at the royal palace. Cold-dry, hot-dry, hot-wet went the cycle of chills and sweats, in a progression that usually ended in death or, very occasionally, the remission of the patient. Her husband, the viceroy, was beside himself. As the *calentura*—as the Spanish referred to fevers—increased in ferocity, the court physicians performed one phlebotomy after another. But the bleeding only weakened the condesa further, and it looked as though death was inevitable.

There was just one slim prospect. Eight years earlier the corregidor (governor) of Loja had also been on the point of death from an intermittent fever. On hearing this, a Jesuit missionary had suggested that the corregidor might like to try a native remedy he himself had taken some thirty years previously for a similar fever. The cure was the bark of a tree growing in the Andean rain forests high above Loja, a town five hundred miles from Lima in what is now Ecuador. The Indians had known the remedy for generations as *ayac cara*—literally "bitter bark"—or sometimes *quina quina*, the "bark of barks." Ground into a powder and taken as an infusion, it had the reputation of being a powerful fever treatment, or febrifuge. Intrigued by the missionary's story, the corregidor had tried the bark and was cured. Now, he suggested, the viceroy might like to administer it to his wife, too.

To cut a long and romantic story short, the viceroy took the corregidor's advice, the condesa was cured, and a few years later she made a triumphant return to Spain, where she spread the news of the miraculous "fever bark" far and wide.

It was as if, overnight, someone had discovered a cure for cancer, or for AIDs. By the latter part of the seventeenth century, malaria was the scourge of Europe. From Seville to Rome, large tracts of countryside had been rendered uninhabitable by the miasmas that clung to coastal deltas and swamps. Cardinals feared having to attend Vatican conclaves because of the risk of being struck dead by the "noxious air" wafting up from the Pontine marshes below Rome. And in 1638, and again in 1658, fever epidemics had swept through England at harvest time, giving the country—as one eyewitness put it—the appearance of a public hospital. No wonder, then, that within a few years of the condesa's recovery the Spanish were presiding over a healthy trade in the bark from South America, and advertisements for the "Countess's powder" or "Jesuits' bark"—as the cure became known—were appearing all over Europe.

Today the cure is better known by its pharmaceutical name, quinine. For years, quinine was added to water by British expatriates in India as a precaution against fever. And long after it was replaced as a prophylactic by new, synthetic drugs such as chloroquine, the habit of drinking Indian tonic water flavored with the bitter alkaloid remained.

But while quinine's name endures—without quinine, for instance, Stanley and Livingstone would probably never have penetrated Africa—almost no one today remembers the tree from which it is extracted. That is a pity, for the story of the hunt for red cinchona—as the varieties containing the highest levels of quinine became known—is an epic. Moreover, unlike the legend of the condesa's cure, it has the virtue of being true.

I FIRST HEARD ABOUT CINCHONA while on a journalistic assignment in Zurich. The working title of the article I was researching was "The Biggest Robbery in History." There is no need to go into the details here, merely to say that it involved billions of dollars' worth of American bonds stolen to order by the Mafia, and that the headline, while a bit of a come-on, was technically accurate. The point is that in the evening, after interviewing the Zurich police, I had gone out for a quiet bite to eat; to my alarm, however, all the restaurants were packed, and I had ended up having to share a table with a Swiss botanist. Anticipating a dull evening but eager to break the silence, I flippantly asked him what he considered to be the most interesting plant in botany. To my surprise he launched into the story of the hunt for cinchona. It was a tale replete with characters every bit as interesting as those in my own story. What is more, it described an act of international skulduggery that was far more deserving of my headline. For if the theft of the most valuable plant ever to come out of the Andes—perhaps the most valuable medicinal plant ever to be found anywhere—did not constitute "The Biggest Robbery in History," what did?

On my return to London I decided to delve deeper into the history of malaria and cinchona. The first thing I discovered was that the word *malaria*—from the Italian for "bad air"—is of relatively recent vintage, only entering the English language in the 1740s, when Horace Walpole spoke of "a horrid thing called the mal'aria, that comes to Rome every summer and kills one." Before that, the preferred English term was *ague*. Indeed, many British explorers afflicted with malaria continued to talk of ague well into the nineteenth century.

The second thing I discovered was that the legend of the condesa's cure was almost certainly an invention. The source of the story was a book written in 1663 by a Genoese physician, Sebastiano Bado, who was closely associated with the Jesuits. Bado claimed to have heard the story from an Italian bark trader in Peru. Unfortunately, scholars have since shown that the condesa almost certainly enjoyed rude health throughout her time in Lima and never returned to Spain, dying of a sudden illness, probably yellow fever, before embarking at Cartagena in 1641.

Next, I looked up *cinchona* in the botanical journals. I discovered it was a member of the Rubiaceae (madder) family and a close cousin of coffee. Like coffee, it preferred hot, moist climates and thrived on rich organic soils and well-drained slopes, such as are found on the sheer sides of volcanoes. Like coffee, it also often took the form of a shrub or small tree with ovate leaves and fragrant white flowers. However, there the resemblance ended. In flower, cinchonas give off a smell like lilac and attract humming birds in brilliant hues. Older cinchonas grow as high as forty to fifty feet, and when they are ready to fruit, the undersides of their leaves turn a spectacular shade of crimson, making them stand out from the rest of the rain-forest canopy. Unlike coffee, which grows anywhere from sea level to six thousand feet, the best cinchonas are found between four thousand and ten thousand feet. But within this narrow altitudinal band, cinchonas have been discovered as far north as Mérida in Venezuela and as far south as Santa Cruz in Bolivia—a distance of nearly 1,800 miles. More important, the cinchona is a hybrid. The tree is prone to cross-pollination and comes in so many different varieties that a strict classification is all but impossible.

The inaccessibility of the plant, combined with its botanical complexity, presented a formidable obstacle to early cinchona hunters. But it was only when I began reading about the expeditions to South America in search of the red cinchona—the *cascarilla roja* of Ecuador and *calisaya vera* of Bolivia—that I realized just *how* formidable. To locate specimens whose plants and seeds were worth cultivating, botanists first had to cross the Atlantic and then either sail around Cape Horn or cross the Isthmus of Panama to land on the coast of Ecuador or Peru. From there they had to mount expeditions that would take them up the western slopes of the Andes, through snow-swept mountain passes and over treacherous glaciers. From as high as twelve thousand feet they would then begin the arduous descent to the cloud forests on the eastern slopes of the cordilleras, the Andean mountain ranges, just above the Amazon—a journey that would take them from South American winter to summer in a matter of days. Along the way, they would have to deal with stubborn mules, truculent and untrustworthy guides, and the ever present threat of death from malnutrition, snakebites, and head-hunting Indians. Added to this were the instability of the nascent Andean republics, sudden border closings, and the risk of detection by native bark collectors keen to protect their livelihood from plundering Europeans. Yet in spite of these obstacles there was no shortage of volunteers. Beginning with a French expedition to Quito in 1735, a succession of European botanists arrived in South America in the hope of seizing the bark and making a gift of it to their sovereigns. But to succeed took knowledge, courage, perseverance, and luck. In the absence of one or another of these elements, nearly all the early expeditions in search of red cinchona ended either in tragic accidents or the death of the participants.

The result was confusion. Some of the barks coming out of South America were indeed cinchona, but others were the medically worthless Peruvian balsam tree, known locally as *quina quina* and by Europeans as *quinquina*. To make matters worse, many unscrupulous merchants would stain their bark with aloe to increase its bitterness or substitute cherry bark, which had a similar coloring. When you consider that even genuine cinchona bark could vary widely in purity, color, taste, and weight, it is little wonder that physicians had a hard time preparing effective febrifuges. Depending on the variety of bark used, some febrifuges would contain high levels of quinine, while others would contain none at all. Not only that, for the first two hundred and fifty years after cinchona's discovery, no one had the faintest idea how quinine worked, much less that malaria was a parasite transmitted by mosquitoes.

By the middle of the nineteenth century the world's most powerful nations were becoming desperate. Although by the 1840s bark exports to Europe were worth £1 million a year, British and Dutch demand was fast outstripping supply. A series of disastrous expeditions up the river Niger had underlined the fact that without a regular supply of quinine the British would never succeed in colonizing the malarious interior of Africa. Similarly, the British army in India required 750 tons a year just to prevent its troops from falling victim to the habitual fevers of Bengal and the Western Ghats. The Dutch were similarly anxious to obtain controlled supplies for their malarious colonies in the East Indies, while in the United States the Union's successful blockade of quinine to the Confederacy had created severe shortages of the bark in the malarious southern states, driving home the drug's importance in warfare.

The British and Dutch decided the time had come to seize cinchona plants and seeds and transport them to their plantations in Madras and Java, respectively, where they could raise sufficient quantities to satisfy the demands of empire. Interestingly, this was not presented as theft but philanthropy. By the 1850s the bark forests of northern Peru and Ecuador had been overfarmed, and even in Bolivia the *calisaya* trees were rapidly being stripped to the point of extinction. Clearly it was of urgent importance to "save" this valuable febrifuge for mankind—or so the imperialists argued—and soon they were actively recruiting men with the necessary qualifications and strength of character for the task.

This was the story that attracted me. It was a tale that pitched Europeans against Indians, botanists against buccaneers, and the British against the Dutch, French, and Spanish. But most of all it was a story about the courage and determination of a few men—mostly British naturalist-explorers—who set off into the highlands of Ecuador, Peru, and Bolivia on not much more than a wing and a prayer. Of those explorers three names stood out: Richard Spruce, a hypochondriac Yorkshireman and moss collector who, despite his fear of disease, spent fifteen years wandering the Amazon and the Andes on behalf of the Royal Botanic Gardens at Kew; Charles Ledger, a cockney trader who came to Peru to make his fortune at the age of eighteen and left South America broken and penniless after a series of incredible expeditions through Bolivia and Chile; and Sir Clements Markham, the patron of Robert Scott's two expeditions to the Antarctic and the so-called father of polar exploration, who, in an earlier incarnation as a renowned Inca historian and linguist, had coordinated a series of British-sponsored missions to South America in search of cinchona. The question was how to tell the stories of their epic adventures. Some, like Ledger, had only described their endeavors briefly in letters home, and while Spruce and Markham kept detailed notes and journals, their accounts were often incomplete.

I decided there was only one way to bring their accounts alive, and that was to go where they had gone. I would not be following in their footsteps exactly—that would take years, and besides, many of the borders they crossed are now closed. Instead, I would recall discrete parts of their trails—the ones that led to the heart of their feverish wanderings. In so doing, I would be going not only in pursuit of a plant but also of a riddle. For until the Spanish conquest, South America was almost certainly malaria-free. The conundrum was how cinchona's medicinal properties were first discovered and by whom—the Indians or the Spanish? And if the former, how could they have also known that it would cure malaria, a disease of the Old World, not the New?

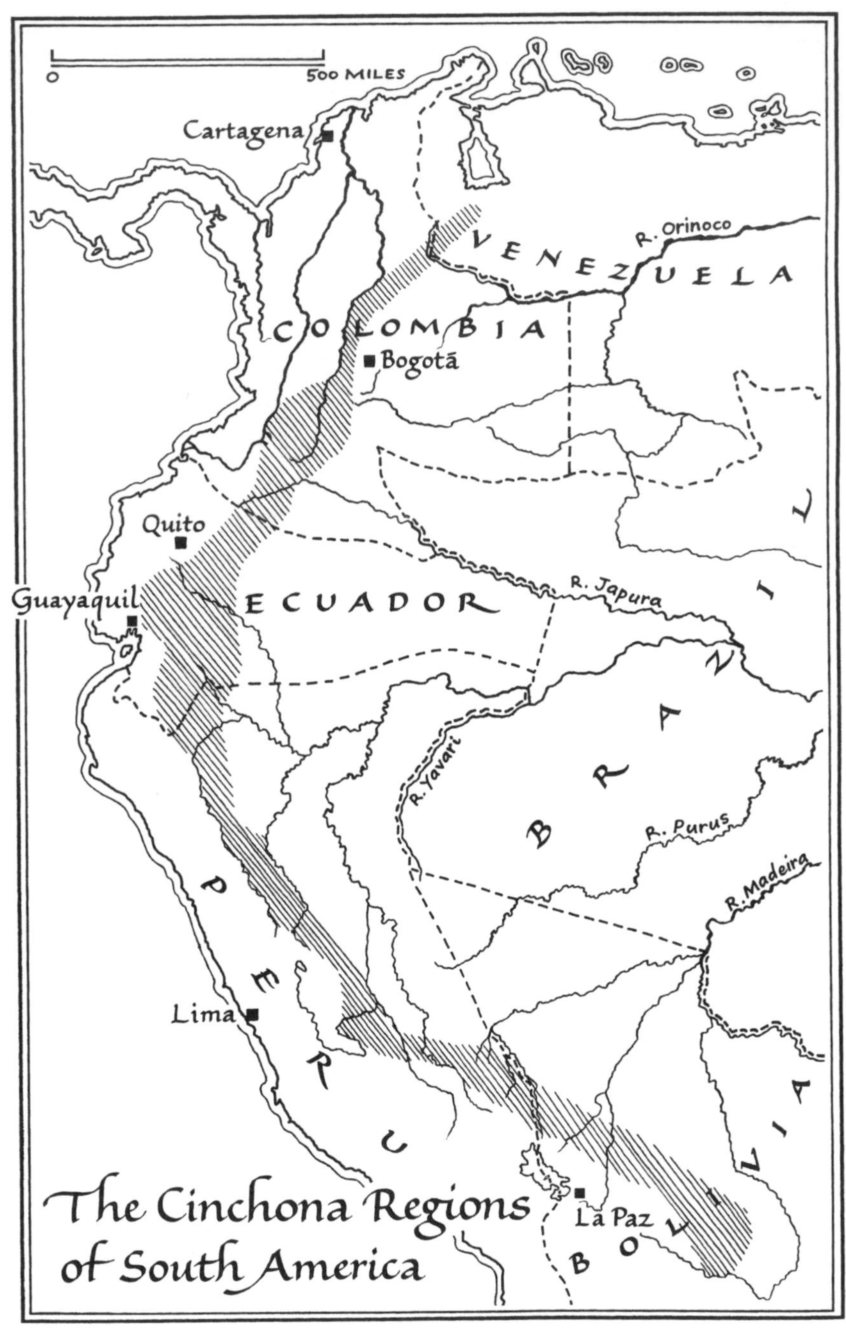
0
500 MILES
Cartagena
VENEZUELA
R. Orinoco
COLOMBIA
Bogotá
Quito
Guayaquil
ECUADOR
R. Japura
BRAZIL
R. Yavari
R. Purus
R. Madeira
PERU
Lima
La Paz
BOLIVIA
The Cinchona Regions of South America

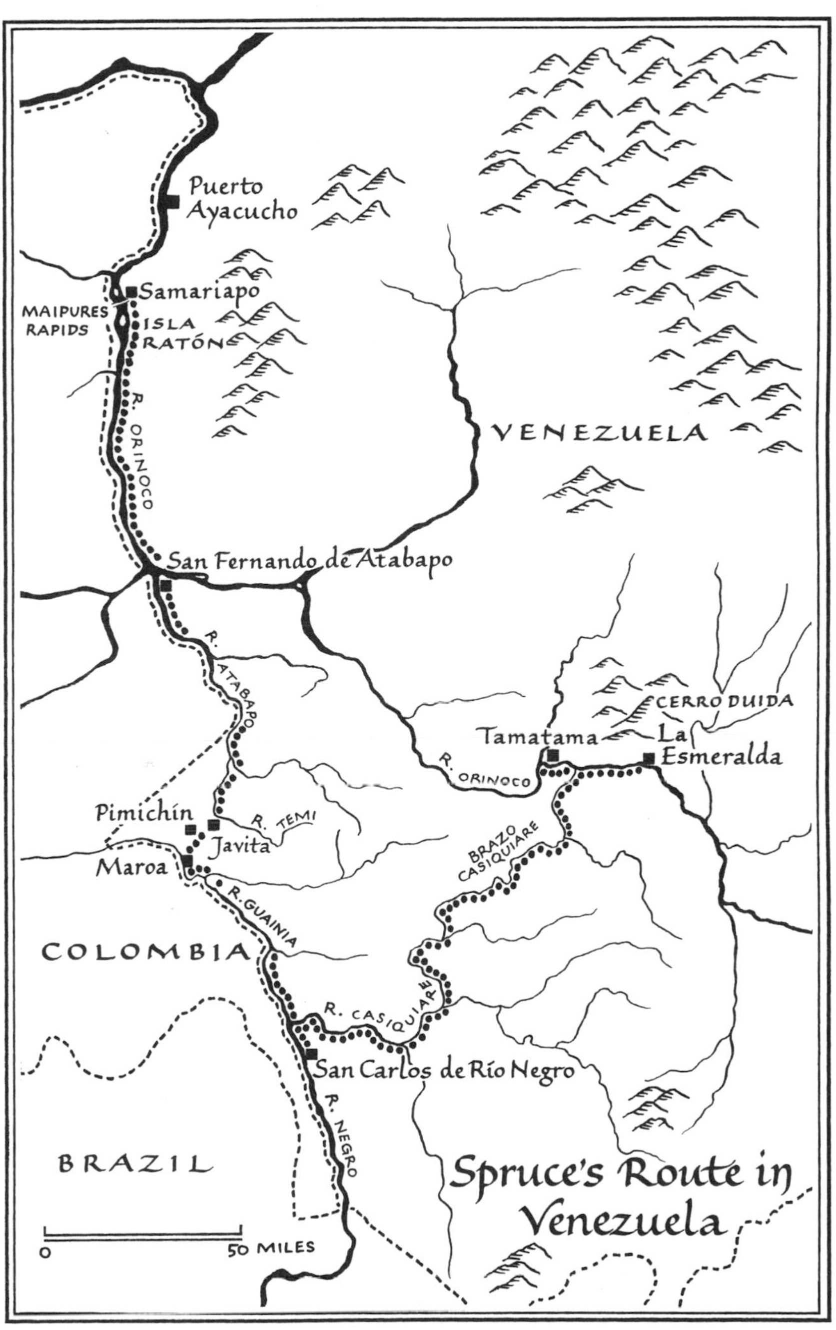

Puerto
Ayacucho
Samariapo
MAIPURES
RAPIDS
ISLA
RATÓN
R. ORINOCO
VENEZUELA
San Fernando de Atabapo
R. ATABAPO
CERRO DUIDA
Tamatama
La
Esmeralda
R. ORINOCO
Pimichín
R. TEMI
Javita
Maroa
BRAZO
CASIQUIARE
R. GUAINIA
COLOMBIA
R. CASIQUIARE
San Carlos de Río Negro
R. NEGRO
BRAZIL
Spruce's Route in
Venezuela
0
50 MILES

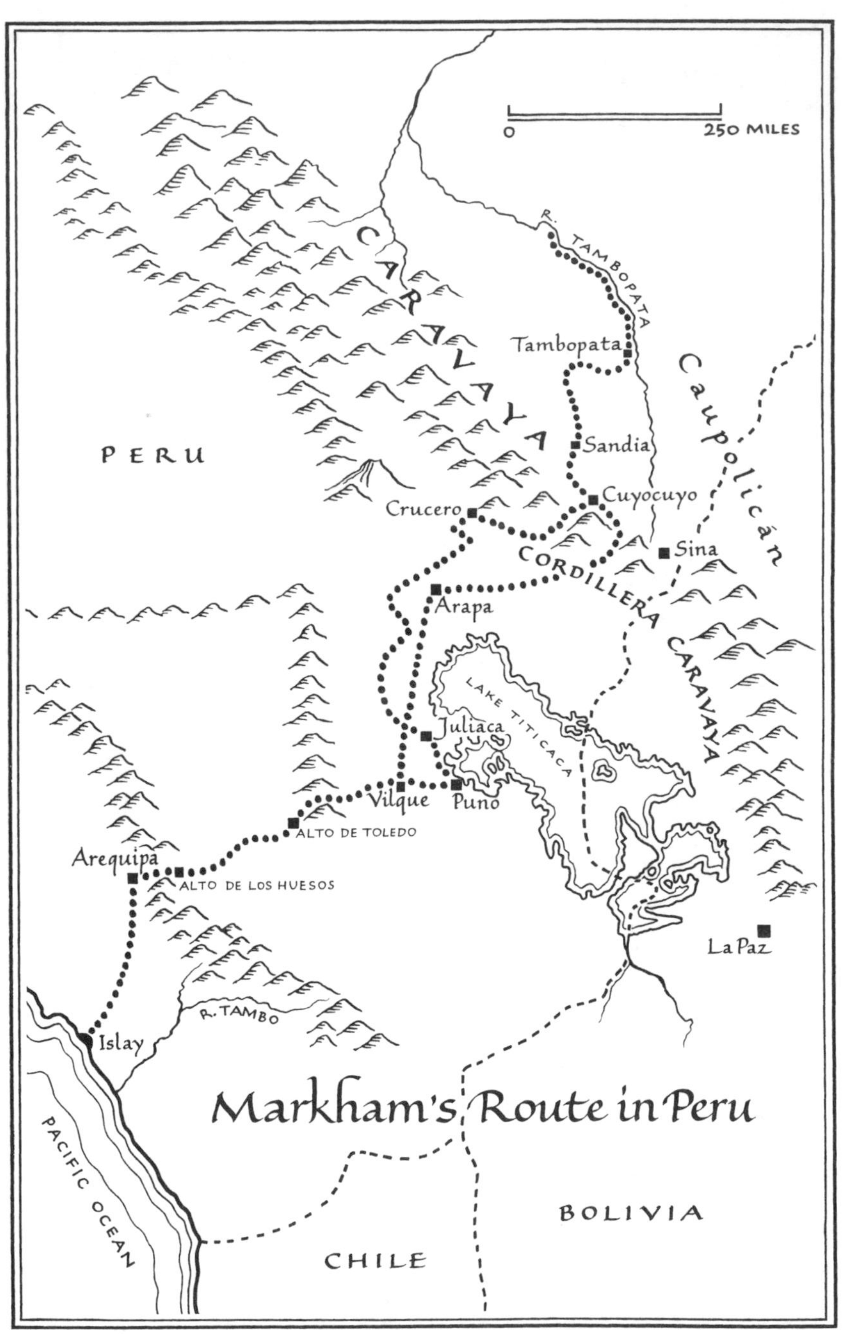
0
250 MILES
PERU
CARAVAYA
R. TAMBOPATA
Tambopata
Caupolicán
Sandia
Cuyocuyo
Crucero
Sina
CORDILLERA CARAVAYA
Arapa
LAKE TITICACA
Juliaca
Vilque
Puno
ALTO DE TOLEDO
Arequipa
ALTO DE LOS HUESOS
La Paz
R. TAMBO
Islay
Markham's Route in Peru
PACIFIC OCEAN
BOLIVIA
CHILE

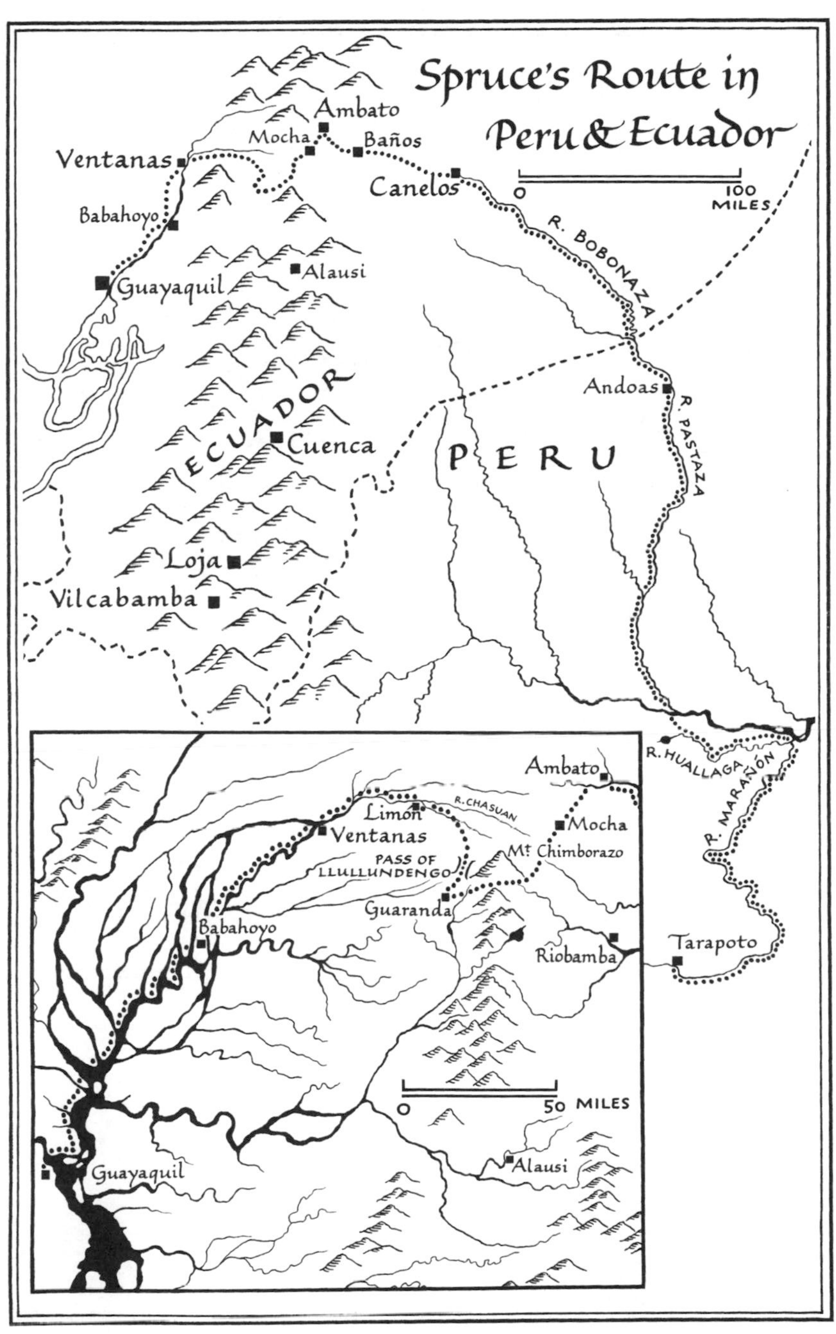

Spruce's Route in Peru & Ecuador
0
100 MILES
Ambato
Mocha
Baños
Ventanas
Canelos
Babahoyo
R. BOBONAZA
Guayaquil
Alausi
Andoas
R. PASTAZA
ECUADOR
Cuenca
PERU
Loja
Vilcabamba
R. HUALLAGA
R. MARAÑON
Tarapoto
Ambato
Limon
R. CHASUAN
Ventanas
Mocha
PASS OF LLULLUNDENGO
Mt Chimborazo
Guaranda
Babahoyo
Riobamba
0
50 MILES
Guayaquil
Alausi

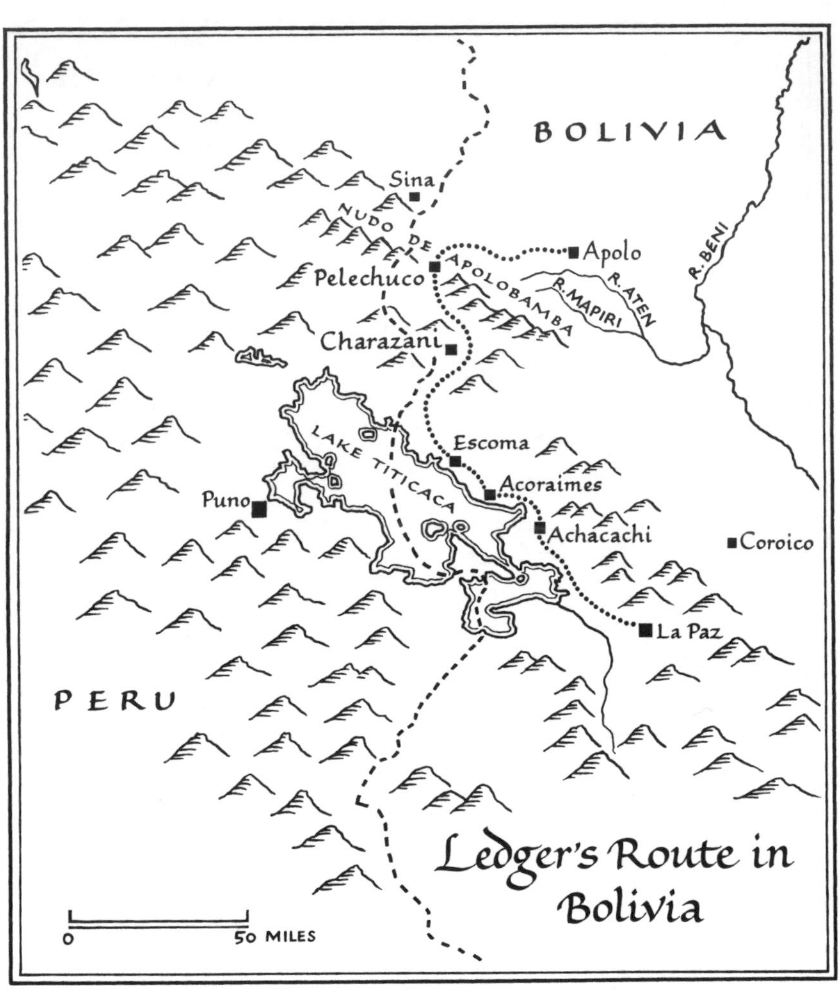
BOLIVIA
Sina
NUDO DE APOLOBAMBA
Apolo
Pelechuco
R. ATEN
R. MAPIRI
R. BENI
Charazani
LAKE TITICACA
Escoma
Acoraimes
Puno
Achacachi
Coroico
La Paz
PERU
Ledger's Route in Bolivia
0
50 MILES

The Fever Trail

1

The Fever

He had a fever when he was in Spain
And when the fit was on him, I did mark
How he did shake.

—SHAKESPEARE,
Julius Caesar

THERE ARE ONLY TWO TOPICS of conversation on the Orinoco during the rainy season: the humidity and the mosquitoes. There are remedies for the former, but there is no escaping the latter. Like the Luftwaffe, the mosquitoes come in waves, each squadron at its appointed hour.

At first light, and foremost throughout the day, are the *jejens*—small blackflies no bigger than pinheads. Their preference is for the foot and ankle, but anywhere will do. In the afternoon, when they swarm from the river, they are so thick in the air it is all you can do to avoid inhaling them. But though the *jejens* leave a rash of tiny swellings punctuated by angry red dots, they are more annoying for their numbers than for the wounds they inflict.

Their deficiency in that department is more than compensated for by their cousin, the horsefly, known as *motuca*. An inky green with white spots, *motucas* are the butchers of the jungle. Unlike the *jejens*, they bite in light or shade so as to afford the weary traveler no respite from their attentions. Fortunately, they are slow, and some satisfaction can be had in swatting them.

The tiny sand fly—*flebotomo*—offers no such opportunity. Almost invisible to the naked eye, they come at twilight just as the campfires are being set and slip through the finest mosquito netting. Like the *jejens*, they leave a rash of small bites, but the irritation is ten times worse.[1]

However, it is the night bombers that are truly to be feared. The Venezuelans call them *zancudos*—a Hispanic word meaning "long-legged" (it is a quirk of Latin American Spanish that the word *mosquito* is reserved for day-biting flies)—but in the English-speaking world they are better known as anophelines, the mosquitoes that transmit malaria. To be precise, in the Alto Orinoco the vector of the malaria parasite is *Anopheles darlingi*. Like other malaria-transmitting anophelines, it has long legs with white socks and a distinctive upward tilt to its abdomen, angled at forty-five degrees when it takes a blood meal. But it is the surgical precision of its proboscis that commands the most attention. When the German geographer Alexander von Humboldt visited the Orinoco in 1800, he complained that of all the river's insect tortures it was the *zancudo*'s "sharp-pointed sucker" that had caused him the "most acute pain." And when Richard Spruce, the English botanical explorer, visited the same waters fifty years later, he wrote that the *zancudo* was capable of penetrating "even a sailor's jackboots."

For once Spruce was exaggerating, but only slightly. *A. darlingi* can penetrate leather and Gore-Tex, and on my first night in the Orinoco it had no trouble biting through the underside of my canvas hammock. Fortunately, I was taking a modern prophylactic—mefloquine—and the next night I slept under mosquito netting. But when Spruce steered his dugout along the Casiquiare Canal to the Orinoco in 1853, he had only one defense against malaria: quinine.

BY 1853, SPRUCE HAD BEEN in South America for four years. He had arrived in Brazil, by steamer from Liverpool, in 1849, gradually working his way up the Amazon to Manaus and from there to the Río Negro. His passage had been arranged by Sir William Hooker, the director of Kew Gardens, and his colleague, George Bentham, and although neither of them knew it at the time, Spruce was to play a crucial role in the quest for cinchona. By the 1850s the Indian government was spending £7,000 a year on imported cinchona bark and £25,000 a year on quinine. In 1852, Dr. Forbes Royle, the reporter on Indian products to the East India Company—the commercial concern that ruled half of India—recommended raising cinchona in the Nilgiri Hills in Madras. "Among the vast majority of medicinal drugs produced in various parts of the world there is not one," wrote Royle, "with probably the single exception of opium, which is more valuable to man than the quinine-yielding cinchonas." Indeed, Royle argued that because of the rising demand for quinine and the seeming inability of the Andean republics to conserve their cinchona forests, there was a real danger that supplies would run out unless the British acted quickly. Because of the difficulty of collecting seeds and the Andean republics' "jealousy" of their bark trade, Royle recommended that the government recruit an experienced botanist who could travel to South America posing as "an agent who had some other objects in view." But rather than send a qualified collector to the Andes, the Foreign Office simply wrote to its consuls in South America, asking them to forward seeds and plants to India. This was easier said than done. Only Walter Cope, the British consul in Quito, succeeded in getting hold of cinchona, but none of the plants survived the voyage to England.

For the moment, none of this was of any concern to Spruce. Although he was first and foremost a bryologist—a student of mosses and liverworts—he had what botanists call the taxonomic eye, and in his voyage through the Brazilian Amazon few new plants escaped his attention. Bentham had agreed to help Spruce finance his travels in South America by selling flora collected in the Amazon to herbaria throughout Europe, and Spruce took every opportunity to add to his botanical presses. In November 1851 he took his piragua—a long canoe made from a hollowed-out tree trunk—from Barra, and by the time he reached the falls of Uanauaca, below São Gabriel, his collection was bulging. With characteristic enthusiasm he wrote to his colleague John Smith at Kew: "Thus far have I advanced into the bowels of the land without impediment, and before adventuring the falls (where I may possibly get a ducking) I seize an opportunity of sending you seed of a beautiful Lythraceous tree which I collected on my way up."

Yet for all the pleasure he took in collecting, Spruce was ill-suited to jungle labor. Born on September 10, 1817, at Ganthorpe, near Castle Howard in the North Riding of Yorkshire, Spruce was plagued from an early age with bronchial trouble. He was an only son whose mother died when he was young, leaving his father, a local schoolmaster, and his stepmother to raise him and his eight stepsisters. Spruce senior taught at Ganthorpe and later Welburn, local schools partially endowed by George Howard, the sixth Earl of Carlisle, and at first it seemed as though Spruce was destined to follow in his father's footsteps. But though he quickly found a position teaching mathematics at Haxby and later at the Collegiate School in York, Spruce never found confinement to the classroom congenial. Inspired by the beauty of the Howardian countryside, he liked nothing better than to spend his spare time wandering the moors and nearby Rye wood, examining local flowers and mosses. His closest friend was a local tinsmith and amateur botanist, Sam Gibson, who kept a copy of Hooker's *British Flora* close by his forge. Encouraged by Gibson, at the age of sixteen Spruce had made a list of all the plants at Ganthorpe, and by nineteen he had completed a flora of the whole Malton district, containing 485 flowering plants.

Thus it was that when the Collegiate School suddenly closed, in 1844, Spruce seized the opportunity to pursue his real passion. Three years earlier he had found a new sedge, *Carex paradoxa*. The discovery had brought him to the attention of Hooker and Bentham, and he now traveled to London to make their acquaintance. The meeting was the catalyst for Spruce's decision to devote the remainder of his life to botany. Bentham suggested he travel to the Pyrenees as a plant collector, funding his expedition by the sale of dried specimens. For someone used to the damp, fogbound Yorkshire hills, the Pyrenees were a revelation. Writing to a friend, the botanist William Borrer, four months after his arrival in the mountains in 1845, Spruce raved: "I can easily conceive why so few mosses have been collected in the Pyrenees, for the flowers are so numerous and so beautiful, that no person, who was not like myself, quite *en tête* of Bryology, would deign to pick up a humble moss!"

Spruce stayed a year, learning Spanish and setting such a pace he often left his guides gasping for breath. The expedition netted five hundred species of mosses, including seventeen that were new to science, and was so significant that Bentham was soon suggesting the more ambitious collecting expedition to South America. Although Spruce would be dependent financially on Bentham's skills as his botanical agent, he had little hesitation in accepting.

Spruce's obsession with his own health presents something of a paradox to would-be biographers. By 1849, for instance, he writes that his bronchitis had worsened to the point where he feared it might be tuberculosis. But given that he was to survive for fifteen years in the Amazon and the Andes, it is hard not to view these and similarly morbid later diagnoses as the conjectures of a pessimistic, if oddly spirited, hypochondriac. The truth is that throughout his time in South America, Spruce was always complaining of catarrh, bilious disorders, gallstones, and other diseases, both real and imagined. When fording rapids or climbing mountains he was prone to sudden coughing fits, and when ascending the Ecuadorian Andes he suffered a sudden and mysterious paralysis that left him unable to sit upright at a desk for the rest of his life.

But what he lacked in physical strength he more than made up for in mental endurance. "I am not one of the forwardest to face perils," he wrote to his Kew colleague John Smith, "but once embarked I think no more of the consequences." Slashing his way through the jungle, he was stung by wasps so often he lost count, and on one occasion he was attacked by vicious *tucundera*—huge black ants—a pain he likened to "that of a hundred thousand nettle stings."

But it wasn't simply his determination that kept him going, it was his curiosity. Whether delayed by the river or the rebellions of his truculent guides or assailed by insects and suspicious natives, Spruce always found something to occupy his orderly mind. In the course of his travels he collected thousands of botanical specimens, including four hundred plants new to science, made the first botanical study of the *Hevea* rubber tree, found the source of the Indian hallucinogen *ayahuasca*, and gathered the largest collection of mosses in the world. In addition, he was a skilled draftsman and linguist, drawing innumerable maps of the regions he explored and collecting the vocabularies of twenty-one Indian languages.

Then there was the "irony" of his malaria. In 1859, while collecting in Ecuador, Spruce was commissioned by the British government to "proceed" to the bark forests and gather cinchona plants and seeds for export to India, with the aim of making quinine freely available to the British Empire. As usual, Spruce did his duty, but had he not been carrying a bottle of quinine when he was on the Orinoco six years previously, he might never have lived to be of service.

LIKE HIS CONTEMPORARIES, Spruce had no inkling of the mosquito's role in transmitting malaria. If he thought about it at all, he would probably have concurred with his friend and colleague the naturalist Alfred Russel Wallace, for whom malaria—like all fevers—was simply the result of noxious gases emanating from decaying vegetable matter in swamps.[2] Indeed, Wallace, who had arrived in South America in 1848, had written to London praising the Amazonian climate, and it was partly as a result of his misleading descriptions that Spruce had booked his own passage a year later. Now that he was there, it mattered little that Wallace's reports about the humidity had been wrong, that his neck, arms, and legs were a mass of suppurating sores, or that, for all his colleagues knew or cared, he could be dead. For Spruce, the Amazon was the realization of a lifelong dream. Like Wordsworth, whose poetry he often turned to in moments of despair, he was drawn by the "sounding cataract . . . the mountain, and the deep and gloomy wood." In his traveling pack he carried a dog-eared copy of Charles Darwin's journals of the voyage of the HMS *Beagle*. Ever since he had first read the book as a youth, he had wanted to do for the botany of South America what Darwin had done for zoology. Now, in his letters home, he could scarcely contain his enthusiasm:

> Fancy if you can two million square miles of forest, uninterrupted save by the streams that traverse it . . . The largest river in the world flows through the largest forest . . . Here our grasses are bamboos sixty or more feet high . . . Our milkworts are stout woody twiners, ascending to the highest tops of trees . . . Instead of

> your periwinkles, we have here handsome trees exuding a most deadly poison. Violets are the size of apple trees . . . Daisies borne on trees like Alders.

If Spruce suffered any dark thoughts amid all this luxuriance, he did not record them, but by now he had plenty of reason to be pessimistic, for the fever was everywhere. In 1851 he had passed through Castanheiro, a notoriously malarious region, so deserted, he wrote in his journal, that the river appeared "dead." Then, at São Gabriel, farther up the Río Negro, he learned that yellow fever had claimed the life of his friend the entomologist Herbert Wallace, who had sailed to South America with him on the brig *Britannia* two years earlier. Now he learned that Herbert's brother, Alfred Russel Wallace, was also seriously ill—this time with malaria—at São Joaquim, a small settlement on the Río Uaupés. "I had sad news two days ago from my friend Wallace," Spruce wrote on December 28, 1851. "He writes me by another hand that he is almost at the point of death from a malignant fever, which has reduced him to such a state of weakness that he cannot rise from his hammock or even feed himself. The person who brought me the letter told me he had taken no nourishment for some days except the juice of oranges and cashews."

Spruce rushed to his bedside, and after taking quinine for four days and nights, Wallace was cured.[3] Incredibly, no sooner had he recovered than he embarked on another expedition, back up the Uaupés and across the Colombian border almost as far as the great falls of Yurupari. At this point, however, his resolve seems to have deserted him, and, exhausted, he sailed back downriver with his collection of animal skeletons and artifacts to Manaus, where he immediately set sail for England. Unfortunately, seven hundred miles from Bermuda his ship caught fire, which consumed the entire collection. It was not the end of his tribulations. Six years later, this time in Indonesia, Wallace would again be seized with a malarial fever during which he would arrive at the theory of natural selection.[4]

WHETHER SPRUCE EXPERIENCED a similarly blinding revelation when eventually he, too, became ill is not recorded. By now he was well acquainted with the insect pests on the Amazon, not only the *jejen* and *motuca*, but the ghastly *pium-piraga*, known as *mosquito colorado* in Spanish because of its distinctive red head, which, Spruce wrote, "sucks an enormous quantity of blood, hanging on till its abdomen is extended to twice or thrice its original size and then falling helpless to the ground." But in passing from the Río Negro to the Casiquiare and Orinoco he found the mosquitoes were more numerous and persistent than anywhere he had previously been. It was Alexander von Humboldt who first remarked on the phenomenon. "In the villages on the banks of the river," he noted in his journal, "the plague of the mosquitoes affords an inexhaustible subject of conversation. When two persons meet in the morning, the first questions they address to each other are: 'How did you find the *zancudos* during the night? How are we today for mosquitoes?' "[5]

Humboldt thought it had something to do with the difference in the color of the water, and he was right. The iodine coloration of the Río Negro is caused by decaying vegetable matter which is thought to release compounds that discourage mosquitoes from breeding. But on the fast-flowing Casiquiare and Orinoco, the water is full of silt and suspended solids that provide the perfect breeding conditions for mosquito larvae.

SPRUCE'S INTRODUCTION to the mosquito plagues of the upper Orinoco came on Christmas Day, at Esmeralda in the shadow of the Cerro Duida, a massive granite mountain that rises 7,861 feet, dominating views from the river. In 1853 the village consisted of just "six miserable huts." Now it is a military outpost with its own airstrip, mission school, and a half-completed museum to Humboldt. But the mud-and-wattle dwellings, though more numerous, are just as miserable, and despite its magnificent aspect the village is no healthier than when Spruce was here. Ever since he'd read Humboldt's description of the cloud-encircled Duida, Spruce had wanted to do what no naturalist had done before—conquer the summit and "rifle its botanical treasures." But no sooner had he reached Esmeralda than he was overcome by a strange melancholia that drained his spirit.

> You will credit me when I say that to the sight Esmeralda is a Paradise—in reality it is an Inferno scarcely habitable by man. When I stood in the middle of the small square, round which are built the houses—the straw doors all carefully closed and looking as if nothing human ever came forth from them—the warm east wind fanning my face and raising the sand in the plaza but bringing no sound of life on its wings—no bird or butterfly to be seen . . . I thought the scene inexpressibly mournful.

Spruce did not have to search far for the answer.

> . . . Throughout the day the very air may be said to be alive with mosquitoes, from which even with closed doors one can only imperfectly escape. I constantly returned from my walk with my hands, feet, neck and face covered with blood, and I found I could nowhere escape these pests. If I climbed the cerros, or buried myself in the forest, or sought the centre of the savannahs, it was the same, but it was worst of all on the river.

Spruce had arranged to meet Don Gregario, the local *comisario* (commissioner), in Esmeralda's main square. The idea was that together they would journey to the as yet undiscovered source of the Orinoco. But there had just been a change of government in Venezuela and no official dared desert his post, for fear of being replaced. Spruce waited for Don Gregario, but after four days, when there was still no sign of him, he could stand the insect torture no longer and "bade adieu to Esmeralda and its mosquitoes."[6] Although it was supposed to be the dry season, it now began to rain heavily, increasing the insect plagues on the Brazo Casiquiare, and after brief detours to explore Lake Vasiva and the Río Pacimoni, Spruce and his Indian porters hurried back to San Carlos.

Then, in May, he headed north. The plan was to explore the Guainía and Atabapo rivers, but the Guainía was navigable by dugout only as far as Tomo. From there, Spruce was forced to hire a small canoe and head for Maroa, where there is a road link to Javita and the Río Temi, which eventually joins the Atabapo. Because of the space restrictions in his canoe and the need for his porters to carry his provisions by foot over Mount Javita, Spruce found himself in a quandary. Should he take two large bundles of papers for his botanical presses, or should he take his medicine chest? He chose the paper. It was very nearly a fatal decision. He set off for San Fernando de Atabapo with only the bare minimum: a small bottle of laudanum, some ipecacuanha root to induce vomiting, and a small vial of quinine sulphate. Emergency supplies.

The town of San Fernando sits on the eastern bank of the Atabapo at the confluence of the Guaviare and the Orinoco. Its broad white-sand beach faces Colombia where, because of the fighting between Marxist guerrillas and the Colombian government, travel is now dangerous. Little else has changed, however. The town still boasts a Plaza Bolívar, a church, and a few stores where vessels replenish their provisions before proceeding north to the cataracts of Maipures. The falls are the magnet. For some seven miles from Atabapo the brown waters of the Guaviare and the lighter waters of the Orinoco run side by side, but as the current quickens they merge until, at Isla Raton, huge granite boulders turn the river into a seething cauldron of dropping channels. The rapids are the end of the road for travelers on the Orinoco. To proceed farther you must disembark at Samariapo and follow a single-track tarmac to Puerto Ayacucho, the capital of Amazonas State, thirty-nine miles to the north.[7] In Spruce's day, however, the disembarkation point was a Salesian mission in the town of Maipures in what is now Colombia.

Spruce left San Fernando on June 7, one month into the rainy season, accompanied by a local trader, Senhor Lauriano, and several Indian guides. On June 18 they stopped at Marana, a small Indian village, and slung their hammocks in a rum distillery swarming with biting ants. They left early the next morning, but owing to Lauriano's crossing the river to sell some goods, it was already dark by the time they reached Maipures. The cataracts were treacherous enough in daylight, and only Lauriano and one of Spruce's Indian porters had previously navigated the narrow creek leading to the mission. Now, with the sun setting, they also had to contend with heavy rain and thunder. Lauriano lit a torch and positioned himself in the prow, allowing the current to pull the boat as he scrutinized the way forward. "Already we could hear the roar of the cataracts mixed with that of thunder and we knew that if we overshot the mouth of the creek the current was so rapid and the fall so near that we were almost certain to be carried away—to certain destruction," Spruce wrote in his journal. Three or four times the rain and wind nearly extinguished the torches. Finally, after several false alarms, Lauriano found the correct channel and steered the canoe swiftly to the mission post.

The town, however, lay three hundred yards farther on, and Spruce and his companions were now forced to cross a marshy savanna traversed by numerous small streams, the classic breeding ground for anophelines. In theory, the chances of contracting malaria from the bite of *Anopheles darlingi* in the Orinoco are low—only one in a hundred carry the parasite—but Spruce was now entering a notoriously malarious area where the concentrations of infectious *zancudos* were much higher. Indeed, Humboldt had been well aware of Maipures's reputation for fevers when he stopped at the mission in 1800. And although he had attributed the fevers to a combination of humidity, bad nutrition, and "pestilent exhalations" from the rocks, he observed that the air was "absolutely filled with venomous insects" and that it was at Maipures that the "suffering from mosquitoes attained its maximum."[8]

As a keen scholar of Humboldt, Spruce cannot have failed to have been aware of these passages, but what could he do? In total darkness now, with the torches guttering, and only the canoe to shield him from the rain, Spruce splashed over the marsh for what "seemed like a mile." When he eventually reached the pueblo he was exhausted, but neither he nor his porters could sleep—the *zancudos* saw to that—and the next night the Indians insisted on sleeping by the river to escape them. Spruce, however, remained in the village. It was the wrong choice.

In Brazil, *Anopheles darlingi* are most active in the early part of the evening, but in the Alto Orinoco they feed continually from dusk until dawn, reaching a peak of activity in the dead of night, when their prey are least suspecting. It is the female who carries the parasite, injecting the protozoan into the blood with her saliva each time she feeds (the male poses no threat, preferring to dine on tree nectar), and like most *Anopheles*, she prefers still, dark spaces—caves or the insides of houses.

The parasite takes four forms—*Plasmodium falciparum*, *P. vivax*, *P. ovale*, and *P. malariae*. Only one, *P. falciparum*, is life threatening (in adults at least), but all are hugely debilitating and, in most cases, difficult to tell apart. The Spanish talk of *el paludismo*, from the Latin word *palus*, meaning "swamp." But the local Yanomami word is far more descriptive. To the Indians who live on the Orinoco, malaria is simply *prisi-prisi*—"the fever."

At first the body displays no symptoms. Having entered the bloodstream, the threadlike malaria parasites (known as sporozoites) make straight for the liver, where they form a cyst and quietly replicate. Two weeks later the cyst suddenly bursts and the young parasites (now known as merozoites) invade the red blood corpuscles. This is the beginning of the fever cycle. Your first clue that something is wrong is a chill, during which you may begin to shiver and your teeth may chatter—the cold-dry phase. Then, as your white blood cells swamp the parasites in an effort to destroy them, your temperature rises dramatically, and the shivering gives way to an extreme sensation of heat. This is the hot-dry phase. Soon you have a raging thirst and headache, and as you slip into delirium, you may believe you are literally burning up. Without warning your headache and temperature subside and you begin to perspire profusely—the hot-wet phase. Even though your bedclothes are now drenched, you feel a huge sense of relief and fall into a grateful sleep, only to wake hours later tired and lethargic.

There follows an intermission, during which you may believe you have recovered. The interval varies according to the malaria parasite. Typically, it lasts twelve to twenty-four hours in the case of falciparum malaria, forty-eight hours in the case of vivax malaria and *ovale*, and seventy-two hours in the case of *malariae*.[9] However, the recovery is an illusion: the cycle of shivers and sweats always returns, with more and more red blood cells being parasitized each time. Viewed through a microscope, the growing malaria parasites are like a corps de ballet: they are all at the liver stage simultaneously, and all burst, as merozoites, from the invaded red cells at the same time. It is beautiful, and potentially deadly.

SPRUCE HAD NO INKLING of the malaria that was probably already coursing through his system. He spent the first few days after his arrival in Maipures exploring the nearby *cerro*, collecting specimens of wild *Barbacaenia*, *Hypnum* mosses, and other species he had not seen before. The greater part of his days, however, was spent turning a side of salted ox in the hot sun so as to stop maggots from breeding in the flesh. With characteristic understatement, Spruce later wrote that this gruesome activity, together with the wetting he had sustained on his arrival, "did not produce any beneficial effect on me."

The first premonition of ague came during the second week of his stay in Maipures. But it was during the four-day voyage back to San Fernando that Spruce writes that the "symptoms declared themselves unmistakably." Spruce's small canoe was so crammed with dried plants and beef flesh that he was forced to sit scrunched up near the entrance to the boat's canopy. Exposed to the sun and rain during the day, and unable to keep warm at night, he was soon "nearly helpless with continued fever." When he reached San Fernando he decided that to proceed farther was folly, and took to his bed.

Spruce made the right decision: judging by his description of his symptoms, he had falciparum, the severest form of malaria. For thirty-eight days he lay in a delirious trance. At first the attacks came every twenty-four hours, but they quickly increased in length and violence, falling into the familiar cycle of thirty-six hours of fever followed by twelve hours of remission.

> Thus at 4–7 pm began the attack, with calafrios and coldness of the extremities, although I had scarcely ever any general shivering of the surface, the cold speedily passing into fever, which continued through the night and through all the next day. Then towards the second night the fever increased in violence—pulse so rapid that it could not be counted—thirst unquenchable—breathing laborious and with great fatigue of chest—mouth constantly filled with viscid saliva, and towards morning the attack would begin to pass off and leave me with forces completely prostrated.

Spruce was now locked in a deadly dance with the parasite. Merozoites have a voracious appetite: in just a few hours they can suck as much as a quarter of a pound of hemoglobin from red blood cells. The result, in the case of a chronic infection, is severe anemia. But what makes falciparum so dangerous is that the still-living blood cells become sticky, which causes them to adhere to one another and to the lining of the capillaries of the brain. This causes the brain to swell and, if left untreated, can result in anoxia and death. The recommended cure then, as now, was quinine. But for some reason Spruce refused to dip into his supply. Perhaps he was uncertain that he really had malaria. After all, the fever could have been the result of any number of tropical diseases. Or perhaps the quinine was simply too precious for him to take without sufficient cause.

The question was very nearly academic. Spruce's porters had traded all his equipment for rum and were so drunk they were unable to attend to his needs. Sustained only by arrowroot broth, he was at risk of starvation and dehydration. A friend had procured a nurse, Carmen Reya, but she also seemed eager to speed him to an early grave. According to Spruce, Reya was a *zamba*, a mixture of black and Indian, "that race by which nine tenths of the most heinous crimes are said to be committed in Venezuela."

> When young she had not been ill-looking, but when out of temper (which for the most part occurred without any reason that I could possibly assign) her face put on a scowl such as one of Macbeth's witches could not outdo. I was already very ill and almost helpless, and nearly all I could do was to ask for what I wanted, yet my slightest word and action was interpreted into something in her disfavour.

Despite these misgivings, Spruce allowed himself to be guided by Reya's advice. During fever fits he drank orange juice, and afterward, instead of replenishing his strength by eating, he took ipecacuanha and another emetic. But gradually he realized that Reya was trying to exacerbate his symptoms. One evening, when he was very sick, she abandoned him for the night, leaving him with just a few jugs of water by his bed. On another occasion she gathered her friends in an adjoining room, taunting him with "all the vile names in which the Spanish language is so rich." According to Spruce her favorite curse was: "Die, you English dog, that we may have a merry *velorio* [wake] with your dollars!"

Spruce was too weak to reply. Instead, he sent for the local *comisario* so that he could give orders for the disposal of his effects. It was only on the fifteenth day of his illness that he became convinced that the continual purges were detrimental to his health and decided to dip into his precious store of quinine. He took 130–195 mg (2–3 grains) once a day. But as his symptoms improved he upped the dose, until by the nineteenth day he was taking 390 mg (6 grains) four times daily. The effect was instant. His fever fell and his appetite returned, and within two days he was able to sit up at the table and drink wine. He continued dosing himself with quinine, coffee, and rum until on August 4 he suffered another attack. The next day, however, he felt better and dined on tapir—a reputed local cure for fevers. "From that day to this," Spruce wrote on September 21, 1854, "I have had, thank God, no more fever."

On August 13—eight days after Spruce's last fever attack—Senhor Antonio Diaz, a Portuguese hammock manufacturer, arrived at Maipures. Spruce knew Diaz, having run into him the previous year at Tomo, where he had commissioned a rare feathered hammock from him as a gift to the Countess of Carlisle, as thanks for the Howard family's support for his botanizing. Made from the thread of the prickly palm and decorated with feathers of hummingbirds, toucans, parrots, and the *gallop ad serve* (cock of the rock), it took fifteen months to complete and is a rare example of local Indian workmanship.[10]

For Spruce it was a fortunate reunion. Anxious to escape his vile nurse, Spruce asked Diaz to take him upriver. Diaz agreed, transferring him to one of his hammocks for the overland journey from Javita to Pinchin. Seven days later, on August 20, Spruce reached Tomo, where he was reunited with his large canoe and, more important, the medicine chest he had left behind in June. He rested for two days to make sure he had recovered his strength, then was on his way again, reaching San Carlos on August 28. Two weeks later he felt well enough to resume his correspondence with Kew. "But a few weeks ago

I thought I would never write to you more," he wrote to George Bentham. "In seeking the wherewithal to sustain life, I came within little of meeting with death . . . Even yet at the end of three months from the first attack, I am very far from having regained my strength, and I am unable to work continuously on anything."

Spruce had had a narrow escape. However, within five years the British government would send him on an even more dangerous mission. It was a task that would test his endurance to the limit. Once again he would be following in the footsteps of Humboldt. But his journey would take him through jungles and over mountain passes that even Humboldt had not dared to tread.

2

The Cure

Amar el árbol es comprender la vida.

—REINALDO ESPINOSA,
Ecuadorian botanist

FROM THE SUMMIT of Mount Cajanuma in southern Ecuador, the province of Loja resembles the surface of a prune: there is not a flat piece of ground anywhere, just a succession of camelbacked ridges and V-shaped valleys rolling toward the horizon. It is as if when God created Loja, He reached down from heaven and gathered the earth into His fist, much as you would clench tight a bedsheet. It has not unfurled yet.

Anywhere else, Loja's daunting topography would be cause for commiseration, but the Lojans are blessed. Although it's just four degrees south of the equator, Loja's altitude and position at the confluence of two weather systems—the Pacific and the Oriente—ensure a perfect year-round climate. Most of the Amazon's heat and humidity never reaches the narrow valley in which the city of Loja nestles but is spilled instead on the eastern slopes of Mount Cajanuma, while to the west the Cordillera de Chilla performs a similar trick with El Niño and other, less violent Pacific fronts. The result is a climate that is neither too hot nor too cold and neither too wet nor too dry, but just right for both plant and animal life.

In Vilcabamba, a subtropical idyll sixty miles south of Loja on the way to the Peruvian frontier, it is said that the inhabitants live to be a hundred and twenty. Few birth certificates are available, but no one who has spent any time in Vilcabamba doubts that the diet and climate are conducive to long life. The terraced slopes groan with yuca, bananas, fresh corn, and San Pedro—a hallucinogenic cactus much in demand on what is left of the hippie trail. Crystal-clear streams—perfect for drinking or bathing in—pour from the verdant hillsides, and you need only walk out of your front door to find oranges, papayas, and avocados the size of bells swinging from the trees. It is thought that when the Incas passed through here on their way to Quito, they were so struck by the cornucopia that they immediately dubbed it a second "sacred valley," after the one they had left behind in northern Peru, and planted *huilca* trees on the hillsides in veneration.

However, it is not for the sacred tree of the Incas but for another plant that the province is best remembered. Cinchona was Loja's first and greatest gift to the world. Anywhere else, that would be an excuse for a whole industry, complete with guided tours and kitsch renderings of the celebrated "fever tree" in ceramic and plastic. But in Loja there is not so much as a museum dedicated to cinchona, let alone a shrine to the Spanish countess after whom it was named. The only clue to Loja's place in history lies three miles south of the city on the road to Vilcabamba. Blink and you will miss it: a crumbling wall, a rusty gate, an overgrown arch and, above it, a simple inscription: *Amar el árbol es comprender la vida* ("To love a tree is to understand life").

The quotation comes from Reinaldo Espinosa, the Ecuadorian botanist in whose honor the gardens were opened in 1949. Driven by a conservationist ethic, Espinosa hoped to restore to Loja many of the native plants and trees, such as cinchona, that had disappeared from the surrounding hillsides in the colonial period. But in a province famed for being the "breadbasket" of Ecuador, his efforts came to naught, and today the gardens—home to 839 species of flora—are the only record of Loja's rich botanical heritage.

Espinosa isn't the only botanist memorialized in the gardens. Open the rusty gate, turn left, and there above the ticket office you will notice another lengthy inscription, this time from Carolus Linnaeus, the Swedish father of botany: "What other scientific investigation is as fatiguing or as difficult as the botanist's endeavours? We know not how to explain the extreme delight of this study [but] it is such that our love of plants surpasses even the love we owe to each other."

It was Linnaeus who in 1742 decided that the quinine tree should bear the name of its first and most glamorous patient, Doña Francisca Henriquez de Ribera, the fourth Condesa de Chinchón. Enamored by the romantic story of her cure, his intention no doubt was to immortalize her name taxonomically. Unfortunately, Linnaeus made an elementary spelling mistake, leaving out the first *h*. By the time other scholars noticed, it was too late, and the genus has been known as cinchona ever since.[1]

To be precise, the species that allegedly cured the condesa was *Cinchona officinalis*—*officinalis* being the Latin term for "storehouse of medicine." Few have described it as memorably as the German naturalist-explorer Alexander von Humboldt. "This beautiful tree," he wrote, when he passed through Loja in 1802, "which is adorned with leaves above five inches long and two broad, growing in dense forests, seems always to aspire to rise above its neighbours. As its upper branches wave to and fro in the wind, their red foliage produces a strange and peculiar effect, recognisable from a great distance."

WHEN, SIXTY-SIX YEARS EARLIER, another explorer, Charles-Marie de La Condamine, had arrived in Ecuador, the cinchona tree was the last thing on his mind. A soldier turned mathematician, La Condamine had been sent to Quito in 1735 by the French Academy of Sciences to settle a dispute then exercising the best minds in Europe: Was the earth, as Sir Isaac Newton asserted, an oblate spheroid—a globe flattened at the poles—or did its diameter lengthen from pole to pole, as the Italian-born French astronomer Jacques Cassini maintained? For largely chauvinistic reasons, most Frenchmen sided with Cassini, arguing that the earth lengthened at the poles and was pulled in at the equator, "much as a potbellied man might pull in his girth by taking a few notches in his belt." One of the exceptions was Voltaire, who had recently returned from England a convert to Newton's theory and translated Newton's *Principia* into French. Another was Voltaire's young disciple La Condamine.

With the help of a group of mathematician-scientists, and the permission of Philip V, the king of Spain, La Condamine was dispatched to the New World to measure one degree of latitude at the equator and thus resolve the dispute once and for all. But no sooner had he arrived in Quito and completed his measurements than La Condamine learned that a rival French team had already reached Lapland and settled the argument in Newton's favor. Marooned in Spanish South America, surrounded by suspicious and increasingly hostile natives, La Condamine needed a new purpose, and quickly, if he was to return to France bathed in glory. It did not take him long to find one. First he would collect plants and seeds of the famous "fever tree" from the forests near Loja. Then he would sail down the Amazon to Brazil, just as the Spanish conquistador Francisco de Orellana had done two hundred years before, hop on a boat to France, and present the quinine tree to his sovereign, Louis XIV. His name would forever be linked with the cure for malaria—or so he must have calculated.

La Condamine headed straight for Loja and Cajanuma, the mountain on whose slopes the finest cinchonas in the New World then grew—the so-called Crown barks of Peru, whose antimalarial properties were then in high demand throughout Europe. Since the supposed cure of the Condesa de Chinchón, the Spanish had presided over a lucrative trade in quinine from Cajanuma and other nearby mountains, such as Uritusinga, shipping the bark around Cape Horn to Seville, where it found its way to Madrid and such other malarious European cities as Rome, London, and Antwerp.[2] By the middle of the seventeenth century the powder of the brittle, highly colored bark was a familiar part of the apothecary's arsenal, but no European had ever laid eyes on the tree from which it came. La Condamine's eyewitness account, *Sur l'arbre Quinquina*, was the first, and it caused a sensation.

La Condamine left Quito on January 18, 1737, following the road that passed directly over the sierra to Cuenca—today's Pan-American Highway—reaching Loja two weeks later. On February 3 he was on his way to the summit of Cajanuma, nine thousand feet above the town of Loja, where he passed the night with a local *cascarillero*—an

Indian who made his living stripping the bark from the forest—returning the next morning with specimens of the leaves, flowers, and seeds. La Condamine explained that "quinquina" came in three colors: white, yellow, and red. The preference was for the red bark, which was more bitter than the others, although the yellow was almost as good. At first glance the trees appeared identical, their leaves, flowers, and exterior bark being the same. It was only when you inserted a knife beneath the tree's skin that you could see that the underside of the yellow bark was softer and more brightly colored.

Not only was La Condamine the first to point to the link between the bark's color and taste and its febrifugal content, but by adopting the Quechua *quina quina*—literally "bark of barks"—he ensured *quinine* a place in the English language. But La Condamine could not resist leavening his account with local folklore and half-remembered popular myths. Thus he accepted the legend of the condesa's cure as true, giving 1638 as the year of her recovery, and adding for good measure that the Indians probably discovered the bark's febrifugal qualities after observing pumas, who suffered their own form of intermittent fever, drinking from a lake that had been infused by the trunks of fallen quinine trees. Years later, as in a game of Chinese whispers, the same story would reemerge in subtly different form: the lake was the same, but this time the discovery was attributed to a feverish Spanish soldier who had drunk from the quinine-rich waters.[3] The stories, like that of the condesa's cure, of course, are highly fanciful, and informative only in so far as they reflect the shifting politics and prejudices of the early Spanish chroniclers. In all probability it took years for the Spanish to discover the secret of the bark—and then only as a result of observations by Jesuit physicians working closely with native Indian healers in the field.[4] But even when the Jesuits began proselytizing in quinine's cause the bark's acceptance was by no means assured.

Quinine had to compete with such well-known and time-tried remedies as viper's broth, crab's eyes, and spiders, which—imprisoned in rags and hung around the patient's neck—were thought to act like therapeutic "cocoons," drawing the "poison" of the fever into their own bodies. The result was that any attempt to proceed by the rational observation of quinine's therapeutic effect was met with incredulity if not outright hostility. For the Galenists, all disease was attributable to either wet or cold humors. Thus quotidian fever was caused by an excess of mucus, tertian by too much bile, and quartan by black bile, a supposed secretion of the spleen. To expel this "*materia morbi*," debilitating measures were required—hence the passion for bloodletting and the enthusiastic use of purgatives and emetics. Incredibly, the bark's opponents argued that it "fixed the humors" and thus maintained the symptoms.

We now know that quinine is a powerful antipyretic that lowers the body's temperature and blocks the malaria parasite's ability to invade human red blood cells, thus interrupting the cycle of fevers and sweats. More important, in stopping malaria, it was the first drug in the medical pharmacopoeia to cure a specific illness. As such, it revolutionized the study and treatment of disease, leading to the foundation of chemotherapy and inspiring the study of homeopathy. But that is not how it appeared at the time; "quinquina," or "china china," as it was sometimes known, was simply another New World bark of unproven efficacy. Not only were supplies from Peru erratic but the substitution of spurious barks, soaked with aloe to give a bitter taste, was rife. Add to this the uncertainty over the correct dosage needed to break the cycle of fevers and sweats, and it is hardly surprising that few physicians were willing to stake their reputations on the remedy, or that the few that were preferred to cloak their adherence to the bark in fantastical tales of a beautiful woman's cure in the distant Spanish viceroyalty of Peru.

In fact, the first reference to the tree comes in a long and dense religious chronicle written by Father Antonio de Calancha, a member of the Augustan order of monks, and printed in Lima in 1633. Most of Father Calancha's thousand-folio treatise consists of accounts of the early missionaries' attempts to convert the Incas in the province of Vilcabamba, Peru, and is hard going. The Yale archaeologist Hiram Bingham, the discoverer of Machu Picchu and Vitcos, who trawled through the text seeking clues to the location of the Lost City of the Incas, describes it as "an omnigatherum of fact and pious fancy." In it, Calancha devotes only one tantalizing sentence to the tree: "There is a tree of 'fevers' in the land of Loja, with cinnamon-coloured bark of which the Lojans cast powders which are drunk in the weight of two small coins, and [thereby] cure fevers and tertians; [these powders] have had miraculous effects in Lima."

Calancha never explains who discovered the tree, how he heard about it, or what the "miraculous effects" might be. Elsewhere in his treatise he describes the Incas' expertise with other medicinal plants, but the quinine tree does not appear to be one of them. Perhaps Loja was beyond Calancha's compass. But you would have thought that Lima was well within it and that had the Condesa de Chinchón been one of the beneficiaries of the tree's "miraculous effects," he would have mentioned it.

The second reference comes from another early Spanish chronicler, Father Bernabé Cobo. In his *History of the New World*, published in 1653, Cobo devotes a whole chapter to the "Arbol de Calenturas" (Fever Tree). Like Calancha, he writes that the tree is found in Loja and has a thick cinnamon-colored bark. However, while he adds that the bark is "very bitter" and that the powders must be taken "a little before the chills begin" to be effective, there is no reference to the condesa or any other cure in Lima.

It is not until ten years later, in 1663, that we get the first written account of the popular story. It comes in the book *Anastasis Corticis Peruviae Seu Chinae Chinae Defensio*, by the Italian physician Sebastiano Bado, and purports to be a true story told to him by an Italian merchant, Antonius Bollus, then living in Peru. According to Bado, the condesa had fallen ill in Lima "thirty or forty years before," thus placing her malaria in 1623–33, not—as La Condamine maintains—in 1638. On hearing of the condesa's fever, the governor of Loja had ridden to Lima to inform the viceroy of how he had been miraculously cured of a similar fever some thirty years previously. Unsure whether to believe the governor's story, the viceroy ordered the court physicians to test the bark on various palace hangers-on to make sure it was safe. Only then did he agree to administer it to the condesa. Her recovery was supposedly instantaneous: "When this was learnt in the City," writes Bado, "the people approached the Vicereine by intermediaries, not so much joyfully and congratulatorily, but supplicatingly, begging her to deign to help them, and say, if she would, by what remedy she had at last so marvellously, so quickly, recovered, so that they, who often suffered from precisely this fever, could also provide for themselves."

Relieved at her narrow escape, Bado continues, the condesa ordered large quantities of bark to be sent from Loja and began distributing the cure with her own hands to the sick of Lima and, later, on her return to Spain, to tenants on her husband's estates at Chinchón, near Madrid. The result was that "just as she herself had experienced the generous hands of God in that miraculous remedy, so all the needy who took it marvellously recovered their health . . . and this bark was afterwards called *Countess's powder.*"

As scholars have subsequently shown, Bado's story is poppycock. Doña Francisca Henriquez de Ribera, fourth Condesa de Chinchón, died on January 14, 1641, in Cartagena. In other words, she never returned from the New World to Spain and thus could not have brought cinchona with her to Madrid. The records do not give the cause of her death but suggest that it was from some epidemic disease then sweeping the Colombian port—most likely yellow fever.

Moreover, it is highly unlikely that the condesa suffered malaria during her time in Lima. Don Luis Gerônimo Fernández de Cabrera Bobadilla y Mendoza, the viceroy and fourth Conde de Chinchón, kept a meticulous diary that he himself observed is "so long and prolix that it will be hard to read." Yet nowhere in the copious jottings recorded by his secretary and official diary keeper, Antonio de Suardo, does the viceroy mention his wife's illness. Indeed, in the eleven-year period covered by the diary, there are only two references to her being unwell. On the first occasion, in November 1630, she is suffering an "inflamed" throat for which "the doctors order her to be bled twice." On the second, in October 1636, she has "flux and cough on the lungs" (again, the treatment is phlebotomy).

The conde is a far better candidate for the legend. His diary makes it clear that between May 1631 and October 1638 he suffered a number of illnesses which frequently confined him to his bed. The first reference to an intermittent fever comes in his entry for May 5, 1631, where he mentions an attack causing "great anxiety and grief." Fifteen days later, on May 20, he is again "ill with an attack of chills and fever," and on June 5 the fever returns, "becoming quartan." And so it continues right up to October 1638, by which time the conde is so debilitated by his illness that the condesa, still apparently in rude health, orders a mass and distributes candles and alms for him throughout the city. On November 8, however, the conde suddenly recovers and, according to the diary, "from this day he steadily improved, having in his illness been in great danger of death." What caused his sudden remission he does not say. But had it been the bark of a hitherto unknown tree, you would have thought he would have recorded it. That he did not is damning evidence against the story related by Bado and all subsequent versions of the legend.[5]

The problem then is how the bark's therapeutic properties were discovered and by whom, and whether Bado had other motives for spreading such a fanciful and easily refutable legend. The short answer is that no one knows for sure. Some medical historians are adamant that the Indians of South America must have known about the quinine tree long before the Spanish arrived in Peru, while others argue just as forcefully that the cure can only have been discovered by the Jesuits. The belief that the Indians got there first is based partly on recent ethnobotanical research showing that the rain forest is a virtual pharmacopoeia to the people who live there. To take just two examples, in the 1940s American botanists were amazed to see Indians treating abrasions with *sangre de drago*, dragon's blood, a natural antiseptic that creates a seal around a wound like a liquid bandage, and toothache with *croton*, a plant in the rubber family.

We also know that Pizarro's soldiers were said to have preferred treatment by native doctors to their own, and that in the first decades following the conquest of South America innumerable herbs, resins, and balsams were dispatched by galleon to Spain. Given their acquaintance with the medical properties of other plants, it seems unlikely that the Indians would have overlooked the virtues of cinchona.

Unfortunately, there are a number of problems with this theory. One is the complete absence of any references to cinchona bark in the hieroglyphics and archaeological remains left by the Incas, or in the "medical books" of the great Indian civilizations of North and Central America.[6] Another is that it is widely accepted that falciparum malaria and even other, milder forms of malaria probably did not exist in South America before Columbus. No malaria, runs the argument, no need for the bark. A third is that even if the pre-Columbian Indians had a limited acquaintance with intermittent fevers, their medical beliefs would have prejudiced them against the bitter-tasting bark.

To this day many South American tribes believe, like the ancient Greeks, that illnesses are either hot or cold and dry or wet, and that the remedy should be the reverse of the condition. Thus if a patient suffers from a hot-dry condition, he or she should be given a cold-wet drink. Quinine—runs the argument—is utterly confusing to the Indian way of thinking because at any point in the malarial cycle that it's administered, it temporarily raises the body's temperature before reducing the fever (hence its relevance to homeopathy). Interestingly, one of the first to accept the conclusion that the Indians could not have known about cinchona was Humboldt. As we have seen, it was while traveling through Loja that Humboldt saw his first cinchona tree. More important for the present argument, he also observed that while "agues are extremely common . . . the natives there . . . would die rather than have recourse to cinchona bark, which, together with opiates, they place in the class of poisons exciting mortification. The Indians cure themselves by lemonades, by oleaginous aromatic peel of the small green wild lemon, by infusions of *scoparia dulcis*, and by strong coffee."

Countering this, some ethnobotanists argue that by the early nineteenth century the Spanish had destroyed most of the Indians' culture and traditions. Thus the medical prejudices Humboldt observed were not the Indians' own but an assimilation from Europe. Nevertheless, Humboldt is adamant that it was the Jesuits who discovered the bark. He claims this occurred by chance when they were testing various trees that had been felled in the Loja forest. "There being always medical practitioners among the missionaries, it is said they had tried an infusion of the cinchona in the tertian ague, a complaint which is very common in that part of the country." Humboldt concludes: "This tradition is less improbable than the assertion of European authors . . . who ascribe the discovery to the Indians."

BUT IS THE CLAIM that the Indians knew about the cure really so improbable? La Condamine may have gilded his sighting of the quinine tree with all sorts of fanciful stories, but on one point he was adamant: the Indians *had* known about the cure but had deliberately hidden the secret from the Spanish because of their hatred of the conquistadors. What made him so certain?

The answer, I believe, lies in La Condamine's introduction to his eyewitness account, where he makes a passing reference to Joseph de Jussieu, the French expedition's botanist, who, he says, has furnished him with "various historical and physical" information on the tree. In fact, it would appear that La Condamine relied heavily on de Jussieu, who was already in the process of completing his own, far more thorough study of cinchona when he set off for Loja. Unfortunately for de Jussieu, La Condamine beat him to publication, and by the time he and his manuscript reached Paris in 1771, no one much cared about his insights. This is a pity, for it is de Jussieu who provides the missing link in the story: "The quinquina is a tree of great height which dominates nearly all its cousins," he writes. "Its trunk is straight and bare, and its width often exceeds nine or ten inches." Unlike La Condamine, de Jussieu distinguishes four types of bark: red, yellow, white, and "*rougesse*." Agreeing that the red is the strongest, he claims that "the Spanish don't dare use it, fearing that by the strength of its heat, which they believe to be excessive, it will provoke a burning fever."

This is a crucial insight—one that suggests that the prejudice against the bark was an imported, not a native, tradition. But de Jussieu goes further, adding a few pages later on that "it is certain" that the discovery is owed to the Malacatos Indians, a tribe from the malarious coastal region of Ecuador who had settled in the valley of Vilcabamba. "Since they had endured much suffering from the hot and humid climate and from intermittent fevers," he writes, "they were forced to find a remedy against the *maladie aussi importune*. After experimenting on various plants, they discovered that the bark of quinquina was the last and almost unique remedy against intermittent fevers."

According to de Jussieu, the Malacatos Indians called the cure *yarachucchu carachucchu*, meaning "bark of the tree for the cold of fevers," or simply *ayac cara*, "bitter bark." The Spanish discovered the secret only when a Jesuit was passing through Malacatos and fell ill with fever. Taking pity on him, the local cacique (chief) offered to cure him and went to fetch some bark from the nearby mountains. On his return he prepared an infusion and the missionary recovered. According to de Jussieu, the Jesuit later noticed similar trees growing throughout Peru and sent the bark to Spain. Significantly, de Jussieu makes no reference to the story of the condesa or the cure of the corregidor of Loja, suggesting that these details were incorporated into the legend at a later date.

That de Jussieu is a superior and more reliable chronicler than La Condamine is underscored by his warning against the adoption of the word *quinquina*, which, he points out, is used indiscriminately by the Peruvians to refer both to cinchona and to the medically worthless balsam tree, *Myroxylon peruiferum.* Unfortunately, few people read de Jussieu's sage advice, and for many years the bark from both trees was distributed under the name "quinquina" in Europe, fueling medical skepticism about the efficacy of the cure.

As far as medical scholars have been able to establish, the Jesuits first introduced the bark into Spain and Italy in 1632; in other words, around the time that Bado says the condesa was being treated in Lima. However, the Jesuits only began to experiment systematically with the bark about twelve years later, when it was brought to the attention of the influential Spanish cardinal Juan de Lugo.

Born in Madrid in 1583 of a noble and wealthy family, de Lugo was raised in Seville, where some of his earliest memories were of the arrival of Spanish galleons loaded with gold and exotic plants and spices from the New World. After studying law at Salamanca, de Lugo entered the Society of Jesus in 1603, where he quickly developed a reputation as a brilliant young theologian with an unrivaled command of jurisprudence. In 1620 he moved to Rome, where he taught for many years at the Gregorian University. Then, in 1643, at the age of sixty, he was suddenly elevated to the rank of cardinal. By now the fevers then rampant in the Roman Campagna had seriously undermined his health. Indeed, such was the fear of malaria that when his benefactor, Pope Urban VIII, died, seven months after Cardinal de Lugo's investiture, foreign cardinals would not cross the Campagna to attend the conclave. Fortunately, de Lugo survived, and when an ally of Spain, Innocent X, was elected pope, he found his influence in the Vatican greatly increased.

At this point, the bark was still something of an unknown quantity. In 1642 the first references to cinchona—under the name *pulvis indicus*—had appeared in Belgium in the writings of a public health official in Ghent. The bark may have been shipped to Antwerp by Michael Belga, physician to the Marqués de Mancera, who had succeeded the Conde de Chinchón as viceroy of Peru. But if cinchona was in circulation in northern Europe at this date, it had yet to win the approval of leading physicians, as can be seen from the death of Ferdinand Habsburg, viceroy of the Netherlands, in 1641 from a tertian fever exacerbated by the usual purges and bloodlettings.

By now Bado, first as physician to de Lugo and later as head of Genoa's two city hospitals, had become an enthusiastic convert. More important, Bado tells us, de Lugo himself had taken the bark and been cured after "lying at the point of death" with a fever. Around 1644, spurred by the reports from his missions in the New World and eager to test the claims of the "miraculous" cure, de Lugo did something no doctor had hitherto contemplated—he instructed the pope's physician, Gabriel Fonseca, to subject the remedy to empirical testing. It was a simple idea, but radical at the time: if observation showed that the bark halted the cycle of shivers and sweats, then the church would recommend its use. Fonseca's report must have been favorable, because not long afterward de Lugo began distributing the bark free to the poor of Rome from his house and palace and at the Hospital of Santo Spirito. Streams of pilgrims began flowing to the Eternal City to try the new medicine, where it was known as *pulvis cardinalis* or *pulvis Jesuiticus*—"Jesuits' bark."

Despite de Lugo's ministrations in Rome, however, the remedy was still little known outside Spain and Italy. The event that changed this was the death in 1649 of the father general of the Society of Jesus. At the resultant council in Rome to elect a new head of the order, de Lugo was able to recommend cinchona to the assembled delegates, ensuring that they would carry the message to their missions throughout the world. The meeting also coincided with the arrival in Seville of large, regular shipments of cinchona from Peru, and under the supervision of the Jesuits it was soon being traded on mercantile exchanges, where it acquired the popular name "Peruvian bark." In no time at all demand was at a fever pitch, with everyone from court physicians to the lowliest apothecaries clamoring for supplies of the miracle cure. Interestingly, in a policy mirrored today by the distribution of the latest generation of antimalarial drugs, de Lugo sold the bark to the rich at high prices in order to give it away to the poor.

In northern Europe, however, physicians remained deeply suspicious of a cure so closely associated with the Jesuits and Rome. Indeed, in 1652 cinchona's cause suffered a serious setback when Archduke Leopold of Austria, governor of the Low Countries, fell ill with a quartan fever. His physician, John Jacob Chiflet, treated him with cinchona and he recovered. But when a month later the archduke suffered a relapse, he refused to take more bark and instead ordered Chiflet to write a book, in defiance of the Vatican, warning the public against cinchona. Published in 1653, Chiflet's *Exposure of the Febrifuge Powder from the American World* was warmly received by many of his contemporaries, wedded as they still were to Galenist principles. Such medical prejudices presented a similarly formidable barrier to the bark's acceptance in Puritan England. To be fair, this was partly political expediency on the part of the medical establishment. In 1649, Charles I had been executed and his Catholic wife sent into exile. Now Oliver Cromwell, Lord Protector of the Commonwealth and guardian of the Protestant faith, was determined to banish the last vestiges of Catholicism from English life. His timing could not have been worse. In the autumn of 1638 there had been an outbreak of "harvest ague," and by 1658 the fever epidemic raged so fiercely in southern England that it was said that there were "scarce hands enough to take in the corn." No one was safe, not even in London, where the river at Westminster and Lambeth had the reputation for being particularly malarious. Perhaps that explains why in July, and again in the autumn of 1658, a small advertisement for the bark appeared in the *Mercurius Politicus.* The advertisement spoke of "the excellent Powder known by the name of Jesuit's Powder, which cureth all manner of Agues, Quotidian, Tertian and Quartan." Imported from Antwerp, "it is to be had at the Black-Spred-Eagle over against Black and White Court in the Old Baily, or at the shop of Mr John Crook, at the sign of the Ship in St Paul's Churchyard, a Bookseller, with directions for using of the same."

Ironically, one of those most in need of treatment was the Lord

Protector himself. Born in 1599 in Huntingdon and raised on the Cambridge fens, Cromwell had been a malaria sufferer all his life. He had frequent bouts of fever as a young man, and there is evidence he suffered repeated relapses both during and after the Civil War. In 1647, for instance, during the crisis between the army and parliament, he was taken seriously ill, and following the Scottish campaign in February 1651, he suffered three fever attacks in as many weeks. Three months later, in May, he was stricken with fever again, this time in Edinburgh, leading his valet, Charles Harvey, to conclude that "this [Scots] air is not so suitable to his temper as that of England." Once again Cromwell recovered, in time to subject the royalist forces to another humiliating defeat, this time at Worcester. The next seven years appear to have been malaria-free—instead Cromwell had to endure the agonies of a kidney stone, gout, and a painful boil. But with the hot summer of 1658, fevers raged once again throughout the British Isles, and on August 24, Cromwell fell ill with what his physician, George Bate, termed a bastard tertian ague. On August 27, Cromwell had severe cold and hot fits. Then, on the night of August 30, a tremendous storm swept the country, uprooting many large trees in St. James's Park and blowing the roofs off houses. With Cromwell's condition deteriorating, his political secretary, the usually unflappable John Thurloe, wrote to Henry Cromwell, his son:

> This is the thirteenth day since the ague took him, havinge been sicke a fortnight before of a generalle distemper of the body. It continued a good while to be a tertian ague, and the burninge fits violent. Upon Saturday it fell to a double tertian havinge two fitts in twenty-four hours; one upon the heeles of another, which doe extremely weaken hym and endanger his life; and truly since Saturday Morning he hath scarce been perfectly out of his fitts. The Doctors are yet hopefull that he may struggle through it, though their hopes are mingled with much feare.[7]

The doctors were right to be fearful; within four days, Cromwell was dead. It is not known whether they offered to treat him with cinchona bark. According to popular legend, they did and he refused it because it was a "Popish remedy." Even if he had agreed, it is far from certain that it would have worked. Cromwell had recovered from previous malaria bouts, which suggests that he had the relatively benign vivax strain of the disease. Falciparum, unless imported by sailors returning from Jamaica, was unknown in England. Indeed, Cromwell's post-mortem suggests that the immediate cause of death was probably septicemia brought on by a urinary infection—although malaria, by weakening his immune system, could have been a contributory factor.[8] More likely, Bate and the other physicians attending Cromwell never offered him cinchona, either because of their own religious prejudices or their suspicions, to some extent justified, of the many bogus powders and preparations on the market.

Ironically, it was thanks to the success of one of these supposedly bogus preparations that the bark finally began to win the confidence of the wider medical profession. The catalyst was Robert Talbor, an unqualified quack from Cambridge who had settled on the coast of Essex—a major smuggling center and then one of the most malarious regions of England—so as to be closer to the seat of the disease. There he experimented with different combinations of bark, eventually perfecting his own mysterious mixture of powders, which he advertised as "Talbor's Wonderful Secret." One of his favorite sales pitches was to tell patients that his elixir would help them expel their "ague cake"—the English term for the enlargement of the spleen that follows infection with the malaria parasite (apparently they were to accomplish this remarkable feat through their mouths). In fact, Talbor's success was more likely due to his assiduous observations of his patients, and soon his reputation was such that his fame had spread to London.

The turning point came in 1672, when Charles II was visiting the coastal fort of Sheerness on the Thames estuary in Kent. One of the members of the king's party was a French nobleman who had previously been treated by Talbor in the Essex marshes. When the Frenchman related the story of Talbor's cure to Charles, the king was intrigued and insisted on meeting the so-called quack in person. Impressed with Talbor and eager to discover the secret of his remedy, Charles made him a physician to the royal household and brought him to London. Styling himself a "pyretiatro"—literally, a feverologist—Talbor began practicing in the metropolis, and in no time his cures were the talk of the town. That same year he published a book, *Pyretologia, or a Rational Account of the Cause and Cure of Agues*, prefacing it with the verse:

> The Learned Author in a generous Fit
> T'oblige his Country hath of Agues Writ
> Physicians now shall be reproacht no more
> Nor Essex shake with Agues as before
> Since certain health salutes her sickly shoar.

Talbor boasted that he had developed a secret remedy for tertian and quartan agues consisting of four ingredients, two of which were native to England, the other two from abroad.[9] Disingenuously, he cautioned against Jesuits' powder, warning that while it was an excellent cure when properly administered, it could also lead to convulsions. In fact, the bark was his key ingredient—two of the others being wine and opium—and the basis of his success.

By now, Talbor was not the only person in England experimenting with the bark. Jesuits' powder had also caught the attention of more respected, establishment scientists, such as Thomas Sydenham. Born in 1624, Sydenham had served as a captain under Cromwell during the Civil War—a cause that led him to miss most of his university work—and had later become a doctor. A friend of the philosopher John Locke and a disciple of Francis Bacon, he was a convert to the novel concept in medical circles that the same sort of empirical observations should be applied to disease as to any other physical effect. Sydenham was the first to describe scarlet fever and added much to the knowledge of gout, from which he suffered. Bringing the same painstaking methods to bear on intermittent fevers, he observed that while the bark did not always cure the ague, it was certainly the only remedy he had seen which managed to halt its progress. In 1666, in his *Methodus Curandi Febres*, he cautiously recommended use of the bark, but only between paroxysms and not at the height of the attack. By 1676, however, he had a better opinion of the treatment. This was partly a result of further experimentation but also because by now he could not ignore Talbor's growing success. Sydenham now recommended the bark unequivocally for the treatment of quartan fevers and no longer suspected it of exacerbating the cycle of fevers and sweats. In his private letters he was even more confident, writing, "I have had but few trials but I am sure that an ounce of bark, given between the two fits, cures."[10]

Sydenham had to be cautious, not least because rivalry at court between pro-Spanish and anti-Papist factions still made advocacy of the bark dangerous. What changed this was Charles II's recurrent malaria. In 1679, at the very moment that pro-Spanish elements were plotting at court to poison him, Charles summoned Talbor to Windsor Castle to treat his ague. Whether this was because he had received news of the plot or feared being bled to death in the name of science is not known. The point is that Talbor administered his quinine mixture and the king recovered. Next, Charles recommended Talbor to his friend Louis XIV, whose only son, the dauphin, was seriously ill with fever. To the consternation of the French medical profession, Talbor once again worked his magic, and soon *le remède anglais* was the talk of *tout* Paris. Louis was so grateful, he awarded Talbor a royal pension and paid him two thousand louis d'or for his proprietary "secret" on condition that it not be published until after his death. So it was that with the new title of chevalier, Talbor began traveling the courts of Europe, curing among others Louisa Maria, the queen of Spain, and—as he put it—"a large number of lesser personages."

Talbor did not enjoy his fame and fortune for long. In 1681, aged just forty, he died and was buried at Trinity Church, Cambridge, where a family monument describes him as Febrium malleus. The following year Louis published the secret formula with an introduction by his chief surgeon, Nichola de Blegny. The revelation that Talbor had been using Jesuits' bark all along did not go down well with the medical profession, particularly as, according to de Blegny, Talbor had bought up all the best bark to prevent others from discovering his secret, probably via his smuggling contacts in Essex. Nevertheless, the publication of the formula made the case for Jesuits' bark irresistible, and *le remède anglais* was swiftly incorporated into the London pharmacopoeia under the name *cortex Peruianus.* Twenty years later, in his monumental *Therapeutice specialis*, the Italian physician Francesco Torti confirmed through exhaustive observation of fever cases that quinquina—or "china china," as he styled it—was the sole remedy for intermittent fevers, and the popularity of cinchona bark was secured.[11]

ONE EFFECT OF TALBOR'S DECEPTION was to underline the importance of traveling to South America to study the trees from which the best bark was taken. "Some authors make two kinds of quinquina, one which they say is wild and of little value, and another which they think is cultivated," he wrote. "To make a true estimate of their quality, it were fit one should be upon the place where they grow." This was the task that La Condamine had begun, but the French academician's description of *Cinchona officinalis* was only the first stage of what would prove to be a complex botanical and pharmacological riddle.

Broadly speaking, cinchonas are distinguished by their ovate or lanceolate leaves arranged in pairs at right angles to one another and by their pretty terminal flowers and seeds. The problem is that, as in the case of the color of the bark, these morphological characteristics are not sufficient to distinguish one species from another. Neither do they necessarily provide a good indication of the level of quinine in the bark. To complicate matters further, cinchona is a hybrid. In the wild, the trees are apt to cross-pollinate, making it difficult to say where a variety ends and a new species begins.

When you consider that cinchonas tend to be found in the most inaccessible parts of the Andes, in forests cloaked for nine months of the year in impenetrable mist, and that the local names for the trees are often at variance with those assigned to them by botanists, it is little wonder that the result has been taxonomical confusion. To this day some botanists argue that there are twenty-three species of cinchona, while others maintain that there are just fifteen, the remaining eight being simply subspecies or varieties of the others. Nevertheless, this is a vast improvement on the position in 1867, when one botanical collector listed 143 species.

But the greatest challenge facing collectors was not the plant but the land and its people. To reach the cinchona forests on the eastern slopes of the Andes, botanical explorers faced daunting natural obstacles: precipitous mountain passes and impassable rock faces; rain forests thick with hornets, mosquitoes, and vicious biting ants; and rushing torrents that could sweep the earth and rocks from beneath a man's feet. Then there was the risk of ambush by head-hunting Indians or by *cascarilleros*, who were dependent on the bark for their livelihood. Nevertheless, in spite of all this, the Spanish had succeeded in removing other botanical treasures, such as the potato, from the Andes. Why should cinchona be any more formidable?

3

The Curse

Good God, when I consider the melancholy fate of so many of botany's votaries, I am tempted to ask whether men are in their right mind who so desperately risk life and everything else through the love of collecting plants.

—CAROLUS LINNAEUS,
Glory of the Scientist

WHEN LA CONDAMINE ARRIVED in the New World with his party of earth measurers in 1735, there were only two ways to reach the Viceroyalty of Peru. One was to disembark at Cartagena on the Caribbean coast of the New Kingdom of Grenada (Colombia) and raft four hundred miles up the Magdalena River to Santa Fé de Bogotá, from where you could proceed along the spine of the Cordillera Central to Popayán and on to Quito. The other was to sail with the Spanish fleet as far as Puerto Bello, where you could cross the Isthmus of Panama to the Pacific and catch a ship direct to Guayaquil.

As it happened, La Condamine was so keen to see Cartagena—no Frenchman had been inside the famous walled city since its foundation—that he stopped there first before proceeding across the isthmus to the coast of Ecuador. Nevertheless, both routes were equally grueling and, from the point of view of health, equally dangerous. The reason was the mosquito.

No one knows when malaria and other mosquito-borne diseases, notably yellow fever, reached the Americas, but the Condesa de Chinchón was by no means their first victim. Six weeks after sacking Cartagena in 1586, Francis Drake had seen fit to flee the port because of a mysterious disease that visited his guards during *la serena*, or evensong.

"They say and hold very firme opinion," wrote Drake, "that who so is then abroad in the open ayre, shall certainly be infected to the death, not being of the Indian or naturall race of those countrey people: by holding their watch, our men were thus subjected to the infectious ayre."

Although it is impossible to be certain, this reads very much like malaria, which is only transmitted at night, when anophelines take their blood meals. Similar accounts of deadly fevers crop up time and again in the records of the early Spanish explorers. For instance, when Vasco Núñez de Balboa crossed the jungles of Darién to become the first European to see the "southern sea," he lost seven hundred and twenty men to fever. And during his exploration of the Orinoco in 1534, Diego de Ordaz recounts how he lost more than three hundred men "*enfeebled by fever* and other diseases induced by the hot, damp atmosphere of the lower stretches of the river."

The Spanish word for the fevers they encountered in the Caribbean was *calentura*—a catchall term that could apply to a plethora of diseases besides malaria, including cholera, typhoid fever, and amebic dysentery. The accepted view today is that if any of these diseases were malaria—and it is a big *if*—the disease probably arrived with the Spanish conquest. The reason is that the first inhabitants of South America were the Indian descendants of Asiatic peoples who migrated across the ice bridge from Siberia to Alaska and through Central America and the Isthmus of Panama. Malariologists argue that such a migration would almost certainly have resulted in falciparum, and most likely other strains of malaria, too, being "frozen out." And while it is possible that malaria could have been introduced to the coast of Chile by Polynesian peoples crossing the Pacific, such a migration remains unproven.

In contrast, we know that by the time of Columbus's first voyage, vivax malaria was endemic to Seville, Lisbon, and the Tagus River, on whose banks the Conde de Chinchón had his estates. In addition, the holds of the ships that brought the early Spanish and Portuguese settlers to South America contained a natural repository for both yellow fever and the falciparum parasite—namely, the blood of black African slaves.

Unlike South America, Africa has almost certainly been malarious for several million years. Indeed, our apelike ancestor, *Homo erectus*, was probably already carrying *Plasmodium falciparum* in his blood when he emerged from the jungle one million years ago and began his long march across the African savanna to the Fertile Crescent and into Asia. The result is that many black Africans and blacks of African descent enjoy natural immunity against malaria due to genetic traits, such as sickle-cell anemia, that have evolved in response to their long exposure to the disease. And even those who do not enjoy such genetically conditioned immunities often acquire limited immunity as a result of repeated exposure to the disease in childhood.

From the Spanish point of view, this resistance to malaria and other tropical diseases made black Africans the ideal laborers to replace native American Indian populations who, ironically, had been decimated by the importation to the New World of Old World diseases such as smallpox and measles. Indeed, Bartolomé de las Casas, one of the early Spanish chroniclers, observed that white settlers believed that the only way a black would die was if they hanged him.

The result was that as early as 1502 the Spanish Crown gave the governor of Hispaniola permission to import *ladinos*—African slaves who had been two or more years in Spain. And by 1518 the demand for black slaves on Hispaniola was so great that Charles I was persuaded to allow blacks to be imported directly from some of the most malarious regions of Africa—notably, the Gold Coast and the Bight of Benin.

From the plasmodium's point of view this was excellent news, as it created a group of human carriers capable of transporting the disease overseas. Beginning with the slave trade in the sixteenth century, this is exactly what happened: slaves were brought from Africa to the Caribbean and the West Indies, turning previously healthy islands into breeding grounds for malaria and yellow fever.

Once the diseases were established in the Caribbean, trade and the imperatives of conquest rapidly spread them to the South American mainland. The Spanish renegade Pedro de Heredia might well have been the first to introduce malaria and yellow fever to Colombia when he set sail from Santo Domingo with several slaves in 1533 and landed at Santa Marta, where he immediately put them to work as *macheteros*, clearing the underbrush. Soon afterward he founded Cartagena and began importing more black laborers, this time to replace Indians who were reluctant to work in the Spanish gold mines.

As the Spanish thirst for gold intensified, so the demand for slaves increased. In the first two decades of the eighteenth century alone, the English brought seventy-five thousand slaves to Latin America—the majority destined for the gold mines in Chocó, Colombia, and the coast of Ecuador. By the early eighteenth century Cartagena had been transformed into a major center of Caribbean slavery.

The legacy of these successive enforced human migrations can be seen in Cartagena today, not only in the Plaza de Coches, the old slave market, but in the wide variety of skin tones on display in its narrow, cobbled streets. Retracing La Condamine's steps through the old walled city—past the pretty colonial-era houses of the Spanish nobility, with their massive iron-studded doors and wooden balconies reminiscent of Seville—you are presented with every shade of the human palette, from black to white to tobacco.

But if racial mixing is the most obvious legacy of the slave trade, it is not the only one. Just as malaria and other tropical diseases retarded European colonization of Africa, so the slave trade, by bringing malaria and probably yellow fever to South America, acted as a brake on European ambitions to overturn the "fact" of the Spanish conquest.

This lesson was dramatically brought home to me at Cartagena's Museo Naval del Caribe by Iris Londono, a former merchant navy seaman who supplements his meager pension by offering bilingual tours of the museum. Having served on British ships for many years, Londono told me he was proud to speak the Queen's English but even prouder of the fact that *cartageneros* still spoke the King's Spanish—

which he attributed to Don Blas de Leso, a one-armed, one-legged Spanish naval officer, and the mosquito. As we wove through the display cases littered with the detritus of corsairs and cannons recovered from the bottom of Cartagena's bay, Londono warmed to his theme.

On March 15, 1741, he explained, the Spanish, African, and mestizo citizens of Cartagena had awoken to a fearsome sight. Spread out across the bay beneath Cartagena's bastions were fifty British warships accompanied by seventy-four transport and supply vessels—the largest armada ever assembled in anger by one nation against another. The fleet contained six thousand marines, eight thousand infantry, four companies from Jamaica, and four battalions of raw American recruits. In all, a force of 18,760 fighting men.

Led by Brigadier-General Wentworth and the English admiral Sir Edward Vernon—or "Old Grog," as he was affectionately known by his men[1]—the task force's aim was to sail into the bay and sack the city, then the wealthiest on the Spanish Main.

England had been itching for an excuse to invade Spain's colonial possessions for some time. For decades, English buccaneers had been raiding ports and shipping lanes on the Spanish Main, egged on by a British public eager for the latest installment of their piratical adventures while Parliament turned a blind eye. Sometimes the Spanish intercepted them, but more often than not they escaped. Then, in 1739, a Spanish coast guard stopped an English merchant ship allegedly loaded with illegal contraband. The captain of the ship was Robert Jenkins, and his ear—or rather what was left of it—was to go down in history.

Every English schoolboy used to know the story of the War of Jenkins's Ear—how the honest master mariner had protested his innocence to the *guardacosta* only to have the dastardly Spaniard cut off his ear in a fit of pique and then spit in what was left of it, "Go tell your king that I will do the same to him should he dare to act as you have." Whether Jenkins was quite as upright or innocent as he subsequently made out, or the coast guard officer quite so despicable, is unclear. But what is indisputable is that when Jenkins limped back to Britain carrying the remains of his ear in a jar and told his story to Parliament, there was an uproar. Inflamed by English commercial interests, and spurred on by the hawkish Vernon, the prime minister, Robert Walpole, reluctantly revived Cromwell's "Western Design"—the blueprint for the breakup of Spain's colonial possessions. Two armadas were assembled: one to sail around Cape Horn under the command of Admiral George Anson and attack the Spanish in the Pacific; the second, led by Wentworth and Vernon, to assault Cartagena. Once Cartagena had fallen and the task force had knocked the commercial stuffing out of the Spanish empire, the English reasoned, it would be a simple matter to sail up the Magdalena River and take the rest of Colombia, too.

In theory, the only thing that stood between the task force and success was San Felipe de Barajas, the huge fort guarding the land entrance to Cartagena, and Blas de Leso, the veteran Spanish naval officer in charge of the city's defenses. But, as Londono now explained—pausing by a grimy display case marked with numbers delineating the course of the naval engagement—Wentworth and Vernon had forgotten to allow for the bay's formidable outer defenses and, equally important, the mosquito.

After raising the British ensign in front of the walled city, Vernon had skirted Cartagena's ramparts and positioned his fleet in front of Boca Chica—the smaller of the two entrances to Cartagena's bay. Having previously taken Puerto Bello with just six ships, Vernon was probably anticipating a short naval engagement, followed by a triumphant entrance into the city. But de Leso had been warned of the task force's arrival and had reinforced the batteries on Tierra Bomba, the island that guarded the narrow channel, delaying the cocksure admiral's advance for fifteen critical days.

In truth, the decisive factor was not so much the Spanish defenses as Wentworth's indolence and incompetence. Not only were his gunners useless, but he put his camp in line with his batteries, so that shot from San Luis de Bocachica that flew over the English artillery landed among his men. Although Vernon repeatedly urged him to storm the fort and put at end to the confrontation, Wentworth demurred. The delay gave Cartagena's mosquitoes an opportunity to go to work on his troops.

It just so happened that Tierra Bomba was peppered with cisterns in which the local inhabitants kept their drinking water. The cisterns provided the perfect breeding ground not only for *Anopheles* but also for *Aedes aegypti*, the mosquito that carries yellow fever. Unlike malaria, yellow fever is a viral infection that is transmitted rapidly from ship to shore and vice versa. The onset of the disease is sudden—the first fever attack usually occurs within five days—and typically results in black vomiting, hemorrhaging, and yellow discoloration of the eyes and skin. In contrast to malaria, which typically has a two-week latency period and is not always fatal, yellow fever is characterized by a high initial death toll.

From the descriptions left by observers of the battle, it is not clear whether the troops and marines were attacked by yellow fever, malaria, or both. Some observers mention black vomit, which occurs only with yellow fever, but others mention only fever. And since most of the men did not begin displaying symptoms until at least two weeks after the initial attack, some epidemiologists believe that malaria must have been the main cause of death. Whatever vector or vectors were involved, however, the effect on the health of the English force was catastrophic.

On April 5, Wentworth finally gave the order to storm San Luis de Bocachica. The Spanish fled without firing a shot, proving that Vernon had been right in urging him to mount an attack earlier. Nevertheless, at this stage Vernon was still confident of victory. Indeed, no sooner had he entered the bay and landed his marines at Manzanillo than he dispatched a courier to London, announcing the fall of Cartagena.

However, the British had reckoned without disease and the determination of Blas de Leso. Although the army had suffered relatively few casualties in the assault on Tierra Bomba, by April 14 "general sickness" had reduced Wentworth's force to thirty-five hundred, and he was pressing to abandon the attack. Vernon was determined to continue. He peppered Cartagena with shell and grapeshot. Then, on the night of April 20, he launched a frontal assault on San Felipe de Barajas. It was a disaster. The fortress was too strong, and as the British struggled to gain a foothold, the Spanish launched a bloody bayonet charge. By dawn eight hundred English dead lay strewn about the ramparts and some one thousand troops had been taken prisoner.

The force had no choice but to retreat to the harbor. It was now that yellow fever and malaria began to take hold, and by May 7 many of the transports had been converted into hospital ships. According to

Smollett, the naval surgeon, the scene was horrendous. "Nothing was heard but groans, lamentations and the language of despair invoking death to deliver [the men] from their miseries," he recalled. When men died, there were too few hands to prepare the corpses for a proper sea burial. The result was that soon the bay was filled with shark-torn bodies floating on the surface.

The final casualty figures make sober reading. Having made an inglorious retreat from Cartagena, the English decided to redeem their honor by attacking Jamaica and Cuba. But here the depredations of the *Aedes aegypti* and *Anopheles* mosquitoes were as great as before, and by the time the task force limped back to England, the army had lost sixty-five hundred men to fever and disease, and the marines reported only thirty-five hundred men fit for duty.

The humiliating defeat underlined as never before the military necessity of having a prophylactic against malaria and other tropical diseases. But it would be another thirty years before naval surgeons would recognize the importance of Peruvian bark in the prevention and treatment of fevers. The first was James Lind, a Scottish naval surgeon who—interestingly—had made his name treating scurvy-ridden survivors of Anson's expedition with lemon peel. In 1765 Lind found himself in Calcutta, where he treated four hundred cases of malaria with cinchona. As a result of those studies, Lind concluded that the bark was most effective when given in full doses as soon as the disease was diagnosed, and in 1771 he persuaded the Admiralty to introduce cinchona in wine to crews on "ships of war on the Guinea station." By the turn of the century, the practice had also spread to other theaters. In December 1803 Admiral Nelson, on the recommendation of Snipe, physician to the Mediterranean Fleet, directed that "a dose of Peruvian-bark, in a preparation of good sound wine or spirits" be given to sailors in the morning before going ashore in marshy areas, "and the same in the evening on his return on board." By 1808, this regimen had been refined, and it was found in Article 24 of the Instruction to Surgeons:

> The Peruvian Bark with wine having been found useful as a preventive, the Surgeon is, when his ship shall be in such tropical climates, to request the Captain a list of men who are to be sent on

> shore on wooding or watering duty, and to administer to each man, previous to his leaving the ship in the morning, a drachm of Bark in half a gill of sound wine, and he is also to give each man the like quantity of wine after he shall have taken the Bark; and a like quantity of Bark and wine proportioned in the same manner is to be given to each man in the evening of the same day on his return to the ship, particularly observing that the Bark administered for this purpose is always to be given in substance, and not in tincture.

Nevertheless, it was not until 1847 that the Lords of the Admiralty issued a directive allowing surgeons to lay in provisions of the prepared quinine wine "in such quantities as their respective complements may seem to require," and then only as a result of the navy's experience off the coast of West Africa and in the notoriously malarious seas of the Bight of Benin.[2]

ENSCONCED IN ECUADOR, La Condamine had plenty of time to mull over Vernon's defeat. One of the ironies was that Puerto Bello was then a major trading post for cinchona. The bark would come from Paita and other Peruvian ports to the Pacific coast of Panama, from where it would be carried across the isthmus by slaves—a grueling four-day journey. The academicians had had reason to be grateful for this traffic when, shortly after leaving Cartagena, de Jussieu had fallen ill with fever at Puerto Bello and prepared a decoction of the bark to cure himself. If only Vernon had seized the opportunity of provisioning his ships with cinchona when he sacked Puerto Bello, history might have turned out differently.

Now the War of Jenkins's Ear began to have repercussions for La Condamine's work, too. After sending his description of the cinchona tree to the French Academy of Sciences, La Condamine had returned to Quito to continue plotting one degree of latitude at the equator. But following Vernon's assault on Cartagena, the whole Pacific coast was up in arms in expectation of an imminent attack by Admiral Anson. As the Peruvians nervously waited for the English to strike from the south, they turned a wary and increasingly suspicious eye on the foreigners in their midst. For six years the academicians had been measuring, drawing maps, and experimenting with strange instruments. The Frenchmen claimed they were measuring the arc of a meridian, but what was their true purpose?

At least one resident of Quito thought he had the answer: the academicians' collection of triangles and theodolites was really a new method for divining gold. "For what could induce gentlemen," he asked rhetorically, "well born and of high station, to lead such a dismal life, uncouth and uncomfortable, climbing mountains, passing over celestial deserts, looking at the stars, for any other reason than earthly profit?"

When it was learned that La Condamine had built a pyramid near Quito, at the point where he had first measured his baseline, suspicion turned to anger. Anson's attack on Colombia never came, but for the next two years La Condamine found himself embroiled in his own war, the War of the Pyramids. The central issue was whether he had the right to erect a pyramid bearing the French coat of arms on Spanish soil. Did this mean he was laying claim to Peru on behalf of the king of France?

After interminable legal wrangling, it was decided that he was not making any such claim and that the pyramid could stay, on condition that the Bourbon coat of arms be obliterated and the names of La Condamine's Spanish explorer colleagues inserted instead. The decision was a diplomatic triumph for La Condamine, but he was now exhausted and anxious to return to France. First, however, he had to fulfill the second part of the vow he had made when he had visited Loja five years earlier: to gather plants and seeds of cinchona and transplant them to Europe, associating his name forever with the cure for malaria.

As he climbed Mount Cajanuma a second time, La Condamine must have thought his fame was assured. Accompanied this time by two Indian porters, he gathered nine young saplings, together with their seeds, placed the plants in an earthen box, and descended the eastern slopes of the Ecuadorian Andes to Borja. The idea was to paddle down the Amazon to Cayenne, on the coast of French Guiana, where he would place the trees *en dépôt* for transportation to the Jardin des Plantes in Paris. Incredibly, he succeeded in re-creating Orellana's epic journey, but at Belém a wave washed over the boat, carrying away the boxes and plants. "Thus," wrote La Condamine, "I lost them after all the care I had taken during a voyage of more than 1200 leagues."

His attempts to cultivate the seeds fared no better. At Cayenne he had taken them to a plantation run by a community of Jesuits. But the soil and climate were not the same as at Loja, and they failed to germinate. By now La Condamine had had enough of South America and was ready to return home, but his trials were not over. He had to wait five more months in Cayenne before finding passage on a Dutch vessel to France. On the way he was twice attacked, once by a British ship and the second time, oddly enough, by a French.

It is hard to feel sorry for him, however. When he finally arrived in Paris in 1745 there was not a savant in Europe who did not wish to speak with him. In the course of his travels La Condamine had not only made the first description of cinchona but had become the first European to describe the *Hevea* rubber tree, fashioning a bag from the latex to keep his scientific instruments dry in the Amazon. He had also conducted experiments with curare that foreshadowed its later use as an anesthetic, and became a convert to inoculation against smallpox—a practice he had seen the Carmelite friars apply with success in Belém (La Condamine went through life with pockmarked skin, the result of a childhood bout of the disease).

Far more deserving of our sympathy is de Jussieu. Born in Lyon in 1704, he was the fourth son in a family of sixteen. Although he studied medicine in Rheims and Paris, he also excelled at mathematics and engineering—indeed, the other academicians often called for his help in making geodesic calculations. But de Jussieu was first and foremost a botanist, and in 1737, and again in 1739, he had also gone to Loja to study the cinchona tree. Unlike La Condamine, de Jussieu experimented with the bark, preparing infusions and testing them on himself to see if the Spanish were correct to fear the "heat" of the red variety. In 1740 he proved that the Spanish prejudice was nonsense when he took *cascarilla roja* to cure himself of a fever at Quito. After practicing medicine there for several years he set off, via Lima, for the cinchona forests of Bolivia, presumably with the intention of seeing whether the yellow *calisaya* barks were equally efficacious. I say "presumably" because he left no account of his travels. We know from the Spanish records that his wanderings took him to Santa Cruz de la Sierra as well as to the silver-mining center of Potosí, and that while in Bolivia he investigated the cultivation of the coca shrub, becoming the first botanist to record the use of the revered stimulant of the Incas.

His intention had been to continue to Buenos Aires and on to France. By now he had a huge collection of specimens all carefully stored away in boxes. But just as he was about to return home to what should have been a hero's welcome, fate intervened. De Jussieu had entrusted the boxes to an Indian servant with strict instructions that he should watch over them night and day because they contained objects of the greatest value. Believing they must be stuffed with gold—for what else was more highly prized by Europeans?—the servant disappeared with the lot.

The boxes were never recovered. In his despair de Jussieu turned his back on botany and Europe, returning to Lima to set up a medical practice. In his spare time he consoled himself with geometry, the only study which, in his own words, "*le satisfaisant par l'évidence de ses démonstrations.*" By the time he eventually returned to Paris in 1771, his memory was gone. He could recall nothing of the earlier hunt for cinchona in Ecuador or the exquisite varieties of *calisayas* he had collected during his wanderings in Bolivia. All that remained was the unedited manuscript he had written on the cinchona tree in 1737. He had kept it with him the whole time and was still clutching it when he was confined to an institution for the insane in Paris. In 1777, de Jussieu's description was finally published in a concise report in the *Histoire de la Societé Royale de Medicine*. Two years later, on April 11, 1779, he died.

It is tempting to view de Jussieu's and La Condamine's failures as more than bad luck. Indeed, those of a superstitious bent remarked that it was as if the tree was protected by some ancient Indian curse. The Spanish had succeeded in stealing the Incas' gold, but it seemed that those who attempted to remove cinchona—a far more valuable product of the Andes—were doomed. There were good reasons why the attempts to steal cinchona were so fraught—not least the botanical complexity and inaccessibility of the tree, and the huge distances involved in transplanting it to Europe. Indeed, prior to La Condamine there is evidence that the Spanish believed that cinchonas did not produce seeds at all but were sterile. The reason for their mistake was simple: cinchona, like coffee, has a short and unpredictable flowering and fruiting season. The Spanish need only have arrived in the bark forests too early—or too late—and they would have missed the seeds entirely.

Luckily, Charles III, the king of Spain, now had a man in South America well equipped to succeed where the French had failed. He had a heavenly name—José Celestino Mutis—and he had come to the New World expressly to gather cinchonas. Born in Cádiz in 1732, Mutis had studied medicine in Seville and Madrid. There his prowess as a doctor had brought him to the attention of the newly appointed viceroy of New Granada, who asked him to join him in Bogotá as the court physician. Mutis readily agreed, believing that the appointment would allow him plenty of time to indulge his passion for botany and astronomy, but when he reached New Granada in 1760 he found his duties were so onerous that he had little opportunity to stray beyond the confines of the palace. In his frustration he wrote directly to Charles III in 1763, urging him to let him form a botanical expedition to collect cinchona. His plea would prove prophetic.

For many years the Jesuits had been nagging the *cascarilleros* to replace felled or dying trees with five more planted in a quincunx—the sign of the cross. But the bark cutters had ignored the Jesuits' commonsense injunctions, and by the 1760s the forests in Loja where the best trees grew were nearly exhausted. Now, in a letter that anticipated the philanthropic arguments of British cinchona hunters a century later, Mutis reminded Charles that he had a duty to his subjects to secure affordable supplies of the febrifuge. "America," he wrote, "is rich not only in gold, silver, precious stones and other treasures but also in natural products . . . there is quinine, a priceless possession of which Your Majesty is the only owner and which divine Providence has bestowed upon you for the good of mankind. It is indispensable to study the cinchona tree so that only the best kind will be sold to the public at the lowest price."

Mutis received no answer. He sent another petition. Then he sent drawings and dried specimens of cinchona flowers gathered in Loja to Linnaeus. The Swedish botanist replied immediately—they were the first examples he had seen since La Condamine's, twenty-five years previously, and they enabled him to complete his classification.[3] But still Mutis heard nothing from Charles III. The silence was doubly frustrating because Mutis had now discovered another variety of cinchona on his doorstep. One day, while riding through the mountains of Bogotá, he had spotted the characteristic rustling of red leaves. He would later claim that the tree, which he christened *Cinchona bogotensis*, was the first cinchona found north of the equator. In fact, by now two other far more valuable species—*C. lancifolia* and *C. cordifolia*—had also been found in Colombia, but it was certainly true that Mutis was the first to classify them, describing all the new Colombian species in a paper, *El Arcano de la Quina*, written by his nephew, Sinforoso, and published in 1793.

The discoveries were important for two reasons. First, it meant that cinchonas could now be harvested in Colombia and transported straight to Spain via Cartagena, rather than having to come from Peru's Pacific coast around Cape Horn or across the Isthmus of Panama. Second, it raised the intriguing possibility that the geographic distribution of cinchona was far wider than had previously been suspected. If it could grow in the Colombian Andes, what valuable varieties might exist at the southern limits of the same mountain chain?

In an attempt to answer that question Charles III in 1777 sent two botanists—Hipolito Ruiz and José Pavon—to hunt for cinchonas in the forests of central and southern Peru and Chile. Accompanied by the French naturalist Joseph Dombey, the Spanish botanists left Cádiz in November 1777 and arrived at Callao in Peru five months later. For seven years they explored Peru and Chile, penetrating the forests of the central Andes and discovering seven new species of cinchona. By the time they returned to Lima in 1785, they had fifty-three boxes packed with specimens. They dispatched them to Spain on the *San Pedro Alcantara*, then resumed their collecting in the Peruvian Andes. But if Ruiz and Pavon thought they had outwitted the curse, they were wrong. In July 1786, while they were collecting in the forests of Huánuco, the house where they kept their journals and specimens accidentally caught fire. Ruiz reached the hacienda just in time to see the flames engulfing his precious manuscripts. Without regard for his safety, he dived into the inferno and had to be dragged out kicking and screaming by his peons. That was not the end of their misfortunes, however. Six months later, within days of reaching Cádiz, the *San Pedro Alcantara* was wrecked off the coast of Portugal, taking Ruiz and Pavon's remaining boxes of specimens with it.

Mutis was spared a similar fate, but even worse was to befall his followers. In 1783, Charles III finally acknowledged Mutis's entreaties and appointed him chief botanist and astronomer to the "Real Expedición Botánica del Nuevo Reino de Granada." Gathering around him a group of talented young naturalists and artists, he scoured Colombia's mountains and river valleys for new species of cinchona and other flora, eventually amassing a huge collection that was stored in a specially constructed library, the grandly titled Instituto Botánica del Nuevo Reino de Granada (Botanical Institute of the New Kingdom of Granada). By 1791, Mutis's staff consisted of some ten scientists and fifteen artists. The most talented of all was Francisco José de Caldas, who had been born in Popayán, in southern Colombia, in 1770. De Caldas was endlessly inquisitive. A student of botany, he was also adept at mathematics, meteorology, and geography. While crossing the Andes, he constructed his own barometer and sextant, and—ignorant of discoveries already made in Europe—independently arrived at the method for gauging altitude with a boiling-point thermometer. Next, he visited Loja and conducted a thorough survey of the cinchona forests of Ecuador and Colombia, producing a map showing cinchona's geographic distribution. By the time he was finished, he had described twenty-two separate species—far more than anyone hitherto. More important, Caldas kept a careful note of the altitude, climate, and soil conditions in which the trees were found. From this he was able to establish that the most valuable species of cinchona rarely grew below four thousand feet or higher than ten thousand feet. He also noticed that they were most abundant on the eastern slopes of the Andes, where the Amazon and its effluents had their sources, and tended to cluster on the sides of river valleys and in the crevices of mountains rather than on more open ground. The red barks, in particular, seemed to prefer areas of steady year-round rainfall and loamy alluvial soil—such as is found near volcanoes.

These insights would later prove crucial to the success of transplanting cinchona overseas, but at the time it seems they were ignored by Mutis, as Caldas was soon complaining that he was not being afforded the recognition he was due. Worse, Mutis made a fundamental error by claiming that *Cinchona lancifolia* and *C. cordifolia*—respectively, the orange- and yellow-colored Colombian barks—were identical to the valuable red and yellow barks described by La Condamine. In fact, as Caldas must have known, since he had visited Loja and Mutis had not, this was false. Thus it was that when Mutis died in 1808, having published just two short papers on cinchona, Caldas seized the opportunity to put the institute's massive collection in order and correct his master's errors. Unfortunately, he had just begun on this immense task when he was interrupted by a revolt against the Spanish government. Caldas promptly joined the rebels, eventually rising to the rank of captain in Bolívar's army. In between the fighting, he returned every now and then to the institute, but by 1816 the rebels were in retreat and Caldas had been taken prisoner by the Spanish. Tried as a traitor, he pleaded for his life to be spared so that he could finish his taxonomy of cinchona, but his pleas were rejected, and on October 29, 1816, he was executed by firing squad. The result was that for all Spain's efforts, the world's understanding of the relationship between the different species of cinchona and the febrifugal properties of the barks had advanced little since the time of La Condamine and de Jussieu.

By the turn of the century this failure had become a source of irritation not only to pharmacists but also to the military. Advances in chemistry meant that pharmacists were at last in a position to identify and isolate the key alkaloids in the bark—the so-called febrifugal salts. The problem was that they still did not know the correct botanical classification for the barks containing the highest levels of quinine. In addition, the practice of adulterating cinchona bark with "false" quinas to boost profits was widespread among merchants and apothecaries, further undermining medical confidence in the remedy. As Mutis had pointed out in his deposition to Charles III, "Such quina tends to be the remedy in which a doctor vainly trusts in the most desperate moments and such finally is the antidote which a poor sick man buys for greater harm to himself. How sad and lamentable would be the spectacle, which a lively imagination could call forth, of a limitless army of loyal subjects tragically lost and abandoned at the foot of your throne!"

In fact it was not Charles III who had to witness this pitiful spectacle but George III, and, once again, the lesson came at sea.

In 1809, encouraged by victories over the French in Portugal and Austria, the British decided to stage a raid against Napoleon in northern Europe. The site chosen was Walcheren in northern Belgium, an area of low-lying marsh in the Scheldt, the estuary that flows from Holland into the North Sea. But as Lord Chatham, the eldest son of William Pitt, assembled an armada on the south coast of England in preparation for the assault, Napoleon remained unimpressed. "We must stop the English with nothing but fever, which will soon devour them all," he predicted, adding that "in a month the English will be obliged to take to their ships."

Napoleon did not have to wait long to see his prediction dramatically fulfilled. Unlike at Cartagena, the navy had taken the precaution of provisioning its ships with Peruvian bark. But it appears that the navy had bought the wrong type of bark or else that, in the absence of any advice from the Army Medical Board, no one had thought to provide sufficient supplies for the army, too. Whatever the case, Walcheren was a disaster.

With the French fleet anchored at Antwerp, the English landed easily in the Scheldt. But on August 10, as Chatham sought to lay siege to Flushing, Napoleon cut the dikes, and by the middle of August, forty thousand English troops were camped on a flat plain surrounded by pools of stagnant water and swarms of mosquitoes.

The first cases of "Walcheren fever" appeared at the end of the month. By the end of September, deaths on Walcheren and the neighboring island of South Beveland had reached a thousand, and several thousand more troops lay sick on hospital ships just as Napoleon had foreseen. By the end of October the situation was even worse, with deaths running at twenty to thirty a day.

The army was wholly unprepared for the crisis. Regimental surgeons warned the healthy not to sleep in the open air, distributed extra rations of rum, and urged troops to smoke cigars in order to "purify" the atmosphere. But when the Scottish regimental surgeon James McGrigor arrived to take command of the medical disaster, he found men sweating on the open decks of ships, unattended save for pails of bark.

The problem was that the demand for bark far exceeded the supply, and it was only by chance that he was able to purchase fresh stocks from a Yankee clipper that had just docked at Flushing.

By the end of the expedition, more than 4,000 men were dead and another 11,500 lay ill. Although some of the casualties were undoubtedly attributable to other diseases, such as typhoid and cholera, there is little doubt that the main cause of death was malaria. So intense was the public reaction that a committee of the entire House of Commons conducted a two-month investigation into the failure of the administration. The inquiry resulted in the dismissal of the Army Medical Board and a complete reform of the army's medical department.

If the British had any lingering doubts about the military value of cinchona, Walcheren erased them, but, from a clinical point of view, identifying which barks contained active levels of the febrifuge was still very much hit and miss. Following the discovery of cinchona in Colombia, the Spanish had begun exporting bark directly from Cartagena. By 1792, the bark shipped from the Caribbean port had reached 360 tons a year, far surpassing exports from Peru and encouraging the British and Americans to open a contraband trade with Colombia. But while the Colombian gray and orange barks were far cheaper, doctors soon noticed that they were not nearly as effective against intermittent fevers as the more highly valued red and yellow barks. Dramatic confirmation of this came in 1799, when the British intercepted a Spanish frigate sailing from Lima to Cádiz loaded with red Peruvian bark. At first, druggists in London and Ostend were unable to dispose of the cargo, but when doctors began testing the red bark on their patients, they discovered it was much better at clearing agues and the price shot up. Within a few years, every major teaching hospital in London had acknowledged the bark's superiority, prompting widespread demand for it from country practitioners, particularly in malarious areas of England such as Kent and Essex. However, it was only when chemists began isolating the essential "febrifugal salts"—the active alkaloids—in the barks that they could at last distinguish which types were more effective.

The first to succeed was Bernardino Antonio Gomez, a Portuguese naval surgeon. In 1812 he soaked powdered gray bark in alcohol, added caustic potash (potassium hydroxide) to the solution to liberate the bases from the salts, and crystallized out an alkaloid to which he gave the name cinchonino. But the real breakthrough came eight years later in France. Pierre Pelletier and Joseph Caventou were brilliant young pharmacists who had already made a name for themselves in 1817 by isolating emetine, a cure for amebic dysentery, from ipecacuanha—the plant Spruce had taken at Maipures to purge himself, in the belief that it would also rid him of fever. The following year they isolated strychnine. Then, in 1820, they subjected the yellow and red cinchona barks to similar analysis. They discovered that Gomez had gotten it wrong: further precipitation and crystallization in alcohol revealed that the base febrifuge was not cinchonino but two alkaloids that occurred separately or together in different kinds of bark. The first they called cinchonine, the second sulphate of quinine. The red bark, they found, contained both. But the yellow bark was abundant chiefly in quinine.[4]

Pelletier and Caventou's discovery was hugely significant, not only clinically but also commercially. At the time, one pound of bark sold for £1 at auction in Amsterdam and London—a not inconsiderable sum of money. To treat a case of malaria, a patient would need nearly two pounds of bark as a first dose, and one pound a week thereafter—in other words, as much as £5 worth in total. But because of the variations in quality, one patient's £5 investment might not be equivalent to another's. Clearly, the first company to succeed in extracting controlled doses of the key alkaloids would be in a position to reap huge rewards. The race was on, but which alkaloid was the most effective against malaria, and from which variety of cinchona did it originate? The first part of that question was easily answered; it was the second that would prove problematic.

Within a year of Pelletier and Caventou's discovery, a number of French doctors had begun testing the "salts" on their patients. Nearly all of them reached the same conclusion: namely, that quinine sulphate was the most effective. Incredibly, Pelletier and Caventou had not applied for a patent to safeguard their discovery. The news of quinine's superiority quickly spread overseas. In 1823 a Dublin doctor tested quinine sulphate on thirty patients, including four who had not responded to treatment with cinchona bark. The treatment succeeded in every case. Around the same time, American physicians began similar experiments, and by 1825 quinine sulphate was the recommended treatment for intermittent fevers in Philadelphia—the erstwhile center of American medical teaching.

Then came the surprise. When chemists tested the barks again, they discovered that the ones with the very highest quinine content were not the red cinchona barks from Ecuador and Peru but the *calisaya* barks from Bolivia. Confusingly, the bark of these trees was yellow, although they were known as "*roja*" by the *cascarilleros* because of their deep reddish purple leaves. The value of *calisayas* had long been recognized in Madrid, where their bark fetched some of the highest prices, but aside from de Jussieu and the Indians of the Bolivian highlands, no one had ever seen the tree in the wild. In addition, botanists still had to answer the vexed question as to whether the red bark tree of Loja—the *Cinchona officinalis* of La Condamine—was one species or several. But at least botanists now knew roughly where to look. Slowly the mists were beginning to clear, and once again it was a scientific generalist—admittedly the most brilliant of his day—who showed the way.

When Baron Alexander von Humboldt set sail for South America in the summer of 1799 accompanied by the French botanist Aimé Bonpland, he boasted to a friend that he intended to "walk from California to Patagonia." "I'll collect plants and animals, I'll analyse heat and electricity, and the magnetic and electric content of the atmosphere . . . I'll measure mountains, but my true purpose is to investigate the interaction of all the forces of nature." Such a feat was well within his abilities. Part geographer, part naturalist and engineer, Humboldt was said to possess an intellect that encompassed every branch of the known sciences—hence his sobriquet, "the last universal man."

Disembarking at Cumaná, in Venezuela, he and Bonpland were soon about their work. First they investigated the famous "oilbirds" of Caripe, prized by the Indians for their fat. Then they paddled up the Orinoco River in a dugout, gathering plant specimens and conducting bizarre experiments on, among other things, electric eels. Next, having reached Esmeralda—the site of Spruce's travails fifty years later—they retraced their steps and proceeded along the Brazo Casiquiare. It was here that they came face-to-face with the full horror of the biting insects. More important, in navigating the natural canal, Humboldt proved not only that the Orinoco and Amazon river systems were linked, but also that the Brazo Casiquiare flows south across the continental divide into the Río Negro—a hydrological phenomenon that European geographers had previously considered impossible.

Having completed their investigations in Venezuela, Humboldt and Bonpland then set sail for Cuba and, after a long delay, Cartagena, with the intention of circumnavigating the globe. Humboldt's idea was to proceed by sea to Guayaquil and then on to Lima, where he hoped to hitch a lift on a passing French schooner. But hearing that the winds were unfavorable and that the journey might take two or three months, he opted instead for the more arduous overland route via the Magdalena River to Bogotá and from there to Quito. Not only would the journey enable them to pay their respects to Mutis, but also it would take them directly through the cinchona forests of Colombia and Ecuador.

The first leg took eight weeks, and, for all but Humboldt, the journey was an ordeal. Bonpland had only recently got over a bout of typhoid fever on the Orinoco. But now, as they poled slowly up the Magdalena River, he began to feel ill again. This time he had malaria, and as he shivered and sweated it became clear that he was not alone: seven of their rowers were also showing signs of the illness. At Honda, where the Magdalena runs swift and narrow, three of the party were so sick they had to remain behind. But Bonpland struggled on, climbing with Humboldt over the Andes and across the chill savanna of Bogotá at 8,600 feet. Fortunately, they had sent letters ahead, warning Mutis of their imminent arrival, and a party of dignitaries accompanied by the archbishop's six-horse carriage rode out to welcome them.

Mutis—then sixty-nine years old—received the European explorers in his library with great ceremony. Tall and heavyset, with dark, piercing eyes, gray hair, and a series of chins that cascaded over his clerical bands, he cut a majestic figure. By now his fame was celebrated throughout Europe. Linnaeus had bestowed on him the accolade "*Nomen immortale, quod nulla aetas numquam delebit*" (Immortal name, which time cannot obliterate). His herbarium housed twenty thousand plants and a team of thirty draftsmen. Under his direction they had executed six thousand botanical illustrations. However, he was not so famous that he had forgotten how to be a physician, and hearing that Bonpland was ill, his first act was to administer a decoction of one drachm (60 grains, or 3,900 mg) of powdered bark dissolved in half a cup of hot water. Having cured Bonpland's malaria, he then invited his guests to tour his institute. Bonpland and Humboldt were suitably impressed. So impressed, in fact, that they remained in Bogotá examining the collection for five months and later inserted a portrait of Mutis, together with a grateful inscription, in their *Plantes Equinoctiales.* Not until September were they ready to leave for Quito.

The journey took four months and the conditions were appalling. Rain and sleet assaulted them on their climb up through the pass of Quindiu, and on the way down their feet became covered in *niguas*—tiny flies that burrow into the soles and toes. Christmas was spent camping at ten thousand feet. Cold and hungry, Humboldt for once lost his equanimity, commenting that "it would hardly be possible to picture a more horrible road." Then, when they limped into Quito, there was more depressing news: the French schooner they had been hoping to sail in would not be calling at Lima after all.

Geography's loss was botany's gain. Humboldt had no choice but to remain in Ecuador and apply his body and mind to other challenges. The first was to conquer Chimborazo, the "monarch of the Andes," which rises to 20,702 feet—then thought to be the highest mountain in the world. The second was to travel to Loja to examine the tree whose febrifugal bark had proved such a boon to mankind.

Today, only Humboldt's conquest of Chimborazo is remembered. He and Bonpland ascended to 19,286 feet—an achievement not surpassed until the development of Himalayan climbing. But in visiting Loja they also cleared up some important misconceptions about cinchona's taxonomy. First, they showed that the Colombian *Cinchona lancifolia*, described by Mutis in 1793, was a completely different species from the *cascarilla roja* of Loja. Second, they proved that Linnaeus's description of the species he called *Cinchona officinalis* was a composite, based on the sample La Condamine had sent him and another, completely different, specimen that Mutis had sent years later from Colombia. According to Humboldt, the reason for Mutis's mistake was that he had paid insufficient attention to the forms of cinchona's leaves. In fact, as Humboldt observed, these varied considerably even in plants of the same species, the length and shape depending as much on the climate, elevation, and whether the trees grew in isolation or in clumps.

This was an important insight. It was later shown that the color of cinchona leaves varied at different heights even on the same plant, and the shape of the leaves could also be different depending on their position on the same branch. But the most important characteristic of cinchonas—which was only discovered in 1948—is that their flowers are heterostylous, making self-pollination extremely improbable. The result is that hybridization is extremely common. Indeed, it was only by trial and error—principally by comparing the alkaloids in the bark—and further expeditions to the Andes that botanists were eventually able to narrow down the distinctive characteristics of the genus.[5] This process would take a further forty years, and even then it would not be complete. But after Humboldt, botanists could have little doubt about the complexity of the task facing them.

To recap, cinchonas grew thousands of feet above sea level in some of the most inaccessible ranges of the Andes. On one side they were protected by the Amazon, on the other by precipitous mountain passes and near impenetrable rain forest. And to top it all, the trees producing the most valuable bark varied in size, shape, and color, depending on the climate, altitude, and soil and their propensity for cross-pollination. No wonder it was so difficult to work out which were the *true* cinchonas. The Indians could hardly have devised a more fiendish defense if they had tried.

Now there was another complication: politics. Spain's South American empire was breaking up. The nascent Andean republics were at war with each other and themselves. Travel in the highlands of Ecuador, Peru, and Bolivia—treacherous at the best of times—was becoming positively dangerous. Militias under the command of tin-pot generals roamed the countryside, press-ganging men and supplies into their makeshift armies. To complicate botanists' task further, in 1844 Bolivia passed laws establishing a national cinchona bank and banning the export of cinchona without a license. The idea was to protect the country's monopoly by regulating the trade so as to discourage smugglers and the reckless stripping of the forests. But this is not what happened. Instead, the company to which the government granted the monopoly paid the *cascarilleros* such pitifully low prices for their bark that they had even less incentive to practice good husbandry than before. The result was that by the 1850s the cinchona forests were slowly but surely being raped to exhaustion, and no one seemed to be able to stop it.

Curse or no curse, something had to be done. Mutis had first identified the issue when he submitted his request to Charles III for a botanical expedition eighty years previously. Now another warning echoed across the Andes, this time from the summit of Chimborazo. "If the governments in America," wrote Humboldt, "do not attend to the preservation of the quina either by prohibiting the felling of the trees or by obliging the territorial magistrates to enforce the cutters to guard them from destruction, the highly esteemed product of the New World will be swept from the country."

4

The Vision

The blows of the axe revived our strength as by enchantment and very soon we were near that great and magnificent tree which I was seeing for the first time, and which for so long had been the object of my dreams.

—M. DELONDRE,
French quinine hunter,
on reaching the calisaya *forests of northern Bolivia*

IT IS ONE THING TO ATTEMPT to steal a nation's botanical property but quite another to transport it across jungles and oceans to a continent six thousand miles to the east. Rogue waves and jealous Indians are only the beginning of one's problems. To succeed, the cinchona hunters first needed to find a sure way of preserving the young plants during the arduous sea crossing to Europe—a voyage that involved fluctuations in temperatures from well below freezing at Cape Horn to more than 100 degrees Fahrenheit at the equator.

Until the early nineteenth century, the standard method for moving botanical specimens from one continent to another was to pack them in crude wooden boxes fitted with hoops and loops over which a rough canvas was stretched to shield them from the sun. Usually the boxes were lashed to the deck or, if the captain would allow it, in his cabin. But the boxes had to be continually monitored to ensure that the plants had sufficient moisture and were free of mold, and it was usually a toss-up whether or not they made landfall alive.

Nathaniel Bagshaw Ward, a general practitioner from the East End of London, hit on a completely new method of plant transportation in 1829. A member of the Linnaean Society and an amateur naturalist, Ward made his discovery by chance when he buried the chrysalis of a sphinx butterfly in a moist mold at the bottom of a bottle and covered it with a lid. A week later he was surprised to notice that a seedling fern and grass had sprouted in the mold. He observed that during the day, when sunlight heated the mold, the moisture, unable to evaporate, condensed on the side of the glass. Then, toward evening, when the glass cooled, the moisture ran back down the bottle to the mold, maintaining a constant humidity. Incredibly, the fern and grass lived for four years until the lid finally rusted and rainwater got into the bottle while Ward was on holiday.

Ward had stumbled on the principle of a unique specialized environment: during the day, the soil produces moisture through evaporation, and the plants absorb the carbon dioxide and photosynthesize it with the help of the sun; during the night they emit more carbon dioxide and vapor. As long as there are sufficient nutrients in the soil and sufficient light, the conditions are perfect for plant life.

Ward had often spoken with William Hooker and members of the Horticultural Society about the problems of transporting plants to and from the colonies. In 1833 he decided to try an experiment. Taking old wooden boxes and using his bottle as a model, he manufactured two cases with glass sides and tops that could be hermetically sealed. He then filled the cases with ferns and grasses and sent them to Sydney.

The plants arrived in perfect condition. The real test, however, came on the return journey. This time the cases were filled with native Australian ferns and dispatched to London via Cape Horn and Rio de Janeiro. The temperatures on board fluctuated from 100 degrees Fahrenheit at Sydney to 20 degrees at Cape Horn, where the decks were covered with snow, to 120 degrees crossing the equator. Nevertheless on reaching the English Channel the plants, which had been placed on deck and never watered, were perfectly preserved.

News of Ward's invention fired collectors' imaginations. Now any plant or flower, no matter how fragile or exotic, could be transported from the tropics and raised in greenhouses in England. One of the first to exploit the new invention was the Duke of Devonshire. In 1835 his gardener, Joseph Paxton, sent a young man from the duke's estate at Chatsworth to India to gather orchids and *Amherstia nobilis*, a tree famed for its startling blood-red flowers. On his return, the collection caused a sensation, and soon royals and aristocrats throughout Europe were ordering Wardian cases for their own greenhouses.

The real impact, however, was on economic botany. The Horticultural Society was among the first to use the Wardian cases in a systematic way. In 1843, Robert Fortune, a Scottish horticulturalist on secondment from the society's Hothouse Department in west London, sailed for China with orders to bring new plants and seeds to enliven its collections—in particular Chinese orchids, aquatics, and plants with "very handsome flowers." Braving hostile mobs and suspicious customs officials, in 1846 he succeeded in sending eight Wardian cases from Canton packed with the sweetly scented winter-flowering honeysuckle *Lonicera fragrantissima*, lace bark pine, and the graceful bleeding heart *Dicentra spectabilis*.

Two years later he was back in China, this time as an employee of the East India Company. Not satisfied with forcing the Chinese to partake in its opium trade, the company now wanted Fortune to smuggle "the finest varieties of the Tea-plant, as well as native manufacturers and implements, for the Government plantations in the Himalayas." Disguising himself as Chinese, Fortune gathered up his Wardian cases and set off for the Hwuy-chow district, two hundred miles inland from Ningpo. Staying at rudimentary Chinese inns or in Buddhist temples so as not to arouse suspicion, he rapidly amassed a collection of plants from the three main tea-growing districts of China. Sowing 250 seedlings in the cases, Fortune then set off for the Himalayas. The teas sprouted en route, and on arrival 215 of the plants were found to be in perfect condition, laying the foundations of the Assam and Sikkim tea industry.

Interestingly, it was the Dutch—always more commercially minded than the British—who first attempted to use the new technology to solve the fraught problem of transporting cinchona overseas, and coincidentally, their effort came at the very moment that Spruce was lying ill with malaria on the Orinoco.

For many years, Dutch botanists had been urging their government to introduce cinchona to their colonies in Southeast Asia. But it was not until 1852 that Charles Pahud, the Dutch minister of the colonies, was authorized to employ an agent. The man he selected was Justus Charles Hasskarl, a botanist and the former superintendent of the Dutch gardens at Buitenzorg in Java. Armed with twenty Wardian cases, Hasskarl set sail for South America in December 1852 with orders to collect as many valuable species of cinchona as possible. His original instructions were to land at Guayaquil and proceed first to Loja, but Hasskarl changed his mind en route, landing at Lima instead and crossing the cordilleras in May 1853.

His first stop was Uchubamba in northern Peru, where he collected a number of plants of a species he called *Cinchona ovata*. But his real goals were the *calisaya* forests of southern Peru, and in September he made his way south via Cusco to Sandia, a town near the border with Bolivia. So as not to arouse the suspicion of the *cascarilleros*, Hasskarl posed as a nature lover, saying he liked nothing better than wandering in the cinchona forests contemplating the unspoiled beauty of the eastern Andes. Unfortunately, Hasskarl arrived too late in the *calisaya* flowering season and, finding the seed panicles burst, returned to Lima to bide his time until the following year.

Hasskarl was next seen at Sina, a tiny village on the frontier between Peru and Bolivia, in April 1854. This time he had assumed a false name, José Carlos Muller, and was posing as a businessman of German extraction. Presenting Muller's card to the local governor, Hasskarl made a proposal: Would the governor be willing to help him procure a supply of *calisaya vera*—the so-called true cinchona of Bolivia? The governor declined, probably out of fear, for the Bolivians were notoriously protective of their bark trade and willing, if necessary, to defend their monopoly on *calisayas* by force. Instead he introduced Hasskarl to a Bolivian, Clemente Henriquez, a dishonest and unscrupulous local fixer. For a down payment of fifty dollars, Henriquez agreed to procure four hundred *calisaya* plants, with a further one hundred dollars to be paid on delivery thirty days later. But soon after arriving in the Bolivian bark forests, Henriquez was arrested for illegal collecting and fled back to Sina. There he employed a local Indian to gather plants on the Peruvian side of the border while he confined his own collecting to an area near the Bolivian frontier.

In June he presented the fruits of his labor to Hasskarl. These were not impressive, consisting of some *calisayas*, but of the wrong Peruvian variety, and all sorts of other cinchonas besides. Hasskarl, however, did not know this. He paid Henriquez the balance of the money he owed him and set off across the cordillera to Arequipa and to the port of Islay, where he loaded the plants onto a waiting Dutch frigate, the *Prinz Frederik der Nederlander*, and resumed his real identity.

No sooner had he set sail than news of his deception reached the *cascarilleros* in Sandia. Believing their livelihoods were about to be destroyed by the transplantation of the *calisayas* to Java, they threatened to cut off Henriquez's feet, forcing him to flee a second time—this time back to Bolivia.

They need not have worried. Unknown to Hasskarl, Henriquez had taken the precaution of mixing arsenic with the earth in the Wardian cases. The result was that by the time Hasskarl reached Batavia on December 13, only two of the plants were still alive. Luckily for him, he had also obtained some seeds, and when these were planted in Tjbodas, a range of mountains south of Batavia, they germinated. Pahud was so delighted he made Hasskarl Knight of the Netherlands and Lion and Commander of the Order of the Oaken Crown, and nominated him director of the cinchona plantations, which were to be expanded.

Such are the rewards of crime when it is carried out in the name of the state. But fortunately for the Bolivians the Dutch celebrations were premature. Not only was Tjbodas unsuitable for cinchona cultivation, but subsequent analysis of the bark from the plants—named *Cinchona pahudiana* in the minister's honor—showed they contained hardly any quinine at all. Hasskarl's fate was sealed. In 1856, Pahud, now governor-general of Netherlands India, relieved him of his duties and handed the direction of the gardens to a more experienced botanist, Franz Junghuhn.

It had been an audacious attempt by the Dutch, but not for the first time a scheme to steal cinchona from South America had failed because of human error.

FEW MEN HAD THE PERSPICACITY or patience to learn from such blunders, but it just so happened that one such man was in Peru now. What is more, he had run into Muller/Hasskarl at Cojata, a town on the shores of Lake Titicaca, shortly before the Dutch botanist had returned to his rendezvous with Henriquez at Sandia. The man's name was Charles Ledger. Unlike Hasskarl, he was a merchant, not a botanist. But by the time of Hasskarl's arrival in South America, Ledger had already spent seventeen years in Peru and knew more about the famous *calisaya vera*, or yellow bark of northern Bolivia, than any European alive.

Born on March 4, 1818, in Stockwell, south London, Ledger was just six months younger than Spruce. He came from a large, upwardly mobile working-class family. His father was a mercantile broker, and although Ledger described himself as a real cockney, there was French blood in the family. His father's ancestors had been Huguenots—Protestant refugees from religious persecution in France—and when Ledger was a boy, his uncle's house in Smithfield had been a popular meeting place for exiles fleeing the revolution. After apprenticing as a silversmith, Ledger had set sail for Peru in 1836 on the steamship *Swallow*, with the aim of making his fortune from the growing trade in cinchona bark and Peru's other great export, alpaca wool. He worked his own passage, his only possessions being a box of steel pens, a prized novelty that he intended to sell in South America, and a letter of introduction to several British trading firms in Lima. His investment in the pens paid off as soon as he landed at Buenos Aires, where he sold them for a small profit. Then the *Swallow* proceeded to Lima, where Ledger found a job as a clerk at Naylor's, a respectable British merchant house. Englishmen of Ledger's education were rare in Peru at that time, and after learning the Spanish language and Naylor's administrative procedures, he was put in charge of one of its branch offices, in Tacna. Located between Lake Titicaca and the coast, it was one of the firm's most important trading outposts, and it was through here that all the best-quality Bolivian bark and alpaca wool flowed. Ledger quickly realized that to the untutored eye all the barks looked the same reddish brown—to succeed he would have to become as expert as the Indians at telling one from the other. He threw himself into the study. First he learned to distinguish the Bolivian *roja* from the inferior *calisaya* barks from Peru. Next he became adept at selecting only the bark from the *tabla*, or trunk of the tree, ignoring the bark from the small branches, and learned to prepare the *serons*, the hides used to transport the bark in a dry condition to England.

By 1842, Ledger had decided that he had learned as much as he could at Naylor's and set up on his own. He at once recruited a Bolivian Indian, Manuel Incra Mamani, whom he had met the year before at La Paz. Mamani is an enigma. Very little is known about him other than that he was from the highlands of northern Bolivia, was either of Quechuan or Aymara Indian descent, and possessed a phenomenal knowledge of cinchonas. Like the best *cascarilleros*, he could spot the trees simply by surveying the rain-forest canopy. Nothing escaped his attention: a handful of brightly colored leaves carried by the wind into his path, the peculiar luster of a leaf tip—either was sufficient to warrant further investigation. Gradually, during their long trips together into the Peruvian interior—Ledger on his mule and Mamani on foot, chewing coca—the Indian began to impart his deep love and affection for the tree to his master. It was a profound friendship and one that would bestow incalculable benefits on mankind.

Ledger quickly realized that Mamani had a better grasp of cinchona than any European botanist. During one journey he distinguished twenty-nine different varieties, pointing out the ones likely to contain the highest amounts of quinine. He was also deeply religious. Every now and again they would come to a group of five cinchonas, planted according to the Jesuits' instructions in the sign of the cross, and Mamani would fall to his knees to pray for the departed souls of the "*buenos padres*." In this way Ledger's appreciation of the tree began to grow, and it was not long before he too was imbued with respect for the valuable *calisayas*.

By now good-quality bark was becoming increasingly hard to find. The idea behind the Bolivian cinchona bank had been to regulate the supply of bark and so protect this valuable natural resource for the future. But the monopoly measures simply encouraged a black market in *calisaya* bark. Despite Humboldt's warning, by the 1850s the cinchona forests were slowly but surely disappearing. Against this background the Dutch and British began to argue that it was their humanitarian duty to "save" this valuable plant.

Meanwhile, in Peru the Bolivian monopoly had brought the authorities to a different conclusion. Worried that they were not exploiting the full potential of their own forests, the Peruvians invited Ledger to lead a fifty-six-man expedition to locate new supplies of Peruvian *calisaya* in Huánuco, a region north of Carabaya. Although Mamani was skeptical of discovering cinchona there, Ledger was convinced his fortune was made. "I, with many others," he later recalled, "entertained no doubts but that we should succeed in discovering large quantities of the best *calisaya*, making rapid fortunes." Each man put up five hundred silver dollars of his own money, and on March 16, 1845, the expedition started from Puno.

Mamani was right. They did not find one genuine *calisaya*, returning three months later empty-handed. Ledger was bitterly disappointed. But there had been no deaths or accidents, and during the long and often arduous expedition Mamani had given Ledger an important insight into the distribution of cinchona.

The key conversation took place one night on their way back to Puno. Ruminating on the failure of the expedition, Ledger suddenly turned to Mamani and asked: "Do you think we shall find the true bark?" Mamani's reply was instant: "No, Señor, the trees here about do not see the snow-capped mountains."

"I could hardly contain my laughter at the moment," Ledger later recalled in a letter to a friend in London, "but when in bed about an hour after I could not sleep for thinking of Manuel's answer."

The next day Ledger unfurled his map and began examining Mamani's proposition. He was right. The cinchona forests stretched from Bogotá, Colombia, in the north, to Santa Cruz, Bolivia, in the south. But the best red barks were found in Apolo, Sorata, and Coroico—the forests dominated by the peaks of Akamani and Illimani in northern Bolivia—and in Loja, Cuenca, and Alausí, in Ecuador, which were overlooked by the volcano Chimborazo.

It was a crucial discovery, one that would hold the key to the success of future cinchona missions. However, Mamani's insight would not be fully appreciated until years later, when the British and Dutch began testing the bark of different varieties of cinchona collected in South America in their plantations in India and Java. For the moment, botanists were a long way from appreciating the relationship between the quinine content of the bark and the climatic, topographical, and other environmental conditions governing where the best trees grew; the reason, once again, was the curse.

IN 1843 ANOTHER GROUP of scientific explorers set sail from France, under the ensign of King Louis Philippe. The expedition was led by the Comte de Castelnau, an indefatigable traveler and sophisticated man of letters, and included two botanists, the Vicomte d'Osery and Hughes Algernon Weddell. Unlike Ruiz and Pavon, Weddell did not plan on stopping at the borders of Peru. Inspired by Joseph de Jussieu, he would penetrate the cinchona forests of Bolivia and make a thorough study of the trees from which the most valuable bark of all came—the *calisayas*. This was no longer considered folly but a matter of commercial and political necessity.

By the 1840s France had a colonial empire in North and central Africa and Indochina that was also plagued by malaria. Like the British and Dutch, the French wanted to transplant cinchonas to their colonies so that they would be assured of a steady supply of quinine. And, like the Dutch, they had set their sights on the Bolivian *calisaya* barks. Thanks to the advances in pharmacology, chemists now knew that the *calisayas* contained the highest levels of quinine, but besides de Jussieu no qualified European botanist had ever set eyes on the tree. Ignorant of Ledger and Mamani's expertise, Weddell and d'Osery planned to enter Peru and Bolivia, collect as many plants and seeds as possible, and then return to Paris. From there it would be a simple matter to send the *calisayas* to North Africa to be raised in plantations in Algiers—or so they must have calculated.

Sailing from Brest, the expedition stopped briefly at Senegal before landing in June at Rio de Janeiro. They spent two years in Brazil, exploring the forests of the Amazon, and then headed for the Urubamba River in Peru. It was at this point that the first in a series of disasters befell them. Castelnau was leading Weddell and d'Osery through an area inhabited by wild forest Indians when they awoke one morning to find that their guides had deserted them and they were headed toward a dangerous waterfall. Fearing that they were about to be murdered or that their baggage would be swept overboard, Castelnau decided to divide the party. D'Osery was sent to Lima with their precious manuscripts, while Castelnau and Weddell continued downriver. Incredibly, d'Osery reached Lima safely and then set off back to the Amazon to rejoin the expedition, which by now had emerged from the forest unscathed. Unfortunately, the authorities in Lima had supplied him with the wrong sort of guides—four head-hunting Jivaro. Somewhere on the Río Marañón, d'Osery disappeared. He was never seen again.

At first Weddell had better luck. In early 1845 he returned to Brazil with Castelnau and headed for the Mato Grosso. Then, in May, he parted from Castelnau and set off alone for Bolivia via Paraguay. He entered Bolivia in August and, after passing through the country of the Chiquito Indians, traveled south to Tarija. It was here, in January 1846, near the nineteenth parallel, that Weddell established the southernmost limits of the cinchona zone, naming the local species *Cinchona australis*. Next he headed north, via Potosí and Cochabamba, until he reached the Yungas, the chain of subtropical valleys that roll east and north of the Bolivian Andes. Here, Weddell writes, "the different species of cinchona multiplied before my eyes." When he stopped at Enquisivi he got his reward: the tree that yielded the precious *calisaya* bark.

Weddell spent several months in the Yungas, gathering information on the collecting of *Cinchona calisaya*, as he christened the tree. The *cascarilleros*, he observed, worked in teams, the most experienced of whom were known as *diestros*. It was the *diestro*'s job to make a preliminary survey of the forest, penetrating the woods in several different directions to establish where the greatest concentrations of cinchonas were. This took great skill, as *calisayas* are rarely found growing in patches or clumps but are more often found singly, lost in a vast green canopy. "It is by the slightest indications that he recognizes the object of his search," wrote Weddell. "The smallest glint of light peculiar to the leaves of a species, the particular colour of the leaves, the peculiar character of a mass of inflorescence . . . Often dead leaves which he may encounter on the ground indicate the neighbourhood of cinchonas, and if they have been swept to the spot where he finds them by the wind, he knows from what direction they have been blown."

Once the *diestro* had located the most promising area of the forest, the other *cascarilleros* would cut a path to it and begin clearing the ground to make room for a drying hut. The smallest sheets of bark, from the branches or the small trunks, yielded the *canuto*—thin cylindrical pieces that curled up in the sun, hence their colloquial name, "quilled bark." The larger strips, from the trunk of the tree, provided the *tabla* or *plancha*—the slabs of bark that, once dried and pressed in the sun, formed the most valuable material for export. Despite the skill needed to spot the cinchonas, Weddell was hopeful of discovering many more species as he traveled north. But as he passed through the gold-mining area of Mapiri and headed to Apolobamba, near the border with Peru, Weddell was alarmed to discover that the forests had been stripped to such an extent that it was necessary to journey for eight to ten days to find a tree still standing.

Eventually, in 1847, he crossed into Peru. After surveying the forests in the province of Carabaya, he journeyed to the ancient Inca city of Cusco, where he was surprised to meet another Frenchman, a Monsieur Delondre, who was also interested in *calisaya*. Delondre had set up a rudimentary plant for manufacturing quinine using machinery based on Pelletier and Caventou's extraction method. Unfortunately, the Bolivians had just sold their entire quota of bark to a manufacturer in New York, and he had no material to process (the local Peruvian bark, of course, did not contain sufficient quinine to make commercial extraction profitable). Desperate to obtain his own source of *calisaya*, Delondre decided to set up a smuggling operation. But the two men he recruited died, one murdered by Indians, the other a suicide. With nothing further to lose, Delondre agreed to join forces with Weddell, and together they set off for the *calisaya* forests of Santa Ana in Bolivia.

It was an awful journey. First they had to climb over the mountains through sheets of freezing rain; then they plunged down through the humid cloud forest. The only thing that lifted their spirits was the clunk of an ax—a signal from their guides that they had found *calisayas*. "We were exhausted," Delondre later wrote, "but the blows of the axe revived our strength as by enchantment and very soon we were near that great and magnificent tree which I was seeing for the first time, and which for so long had been the object of my dreams. I remained in ecstasies, gazing at its beautiful silvery bark, at its large leaves of a greenish-brownish colour, and its flowers with the sweetish perfume, slightly reminiscent of the lilac."

Weddell's description, published in 1849, in which he listed nineteen different varieties of cinchona, was more sober and scientific.[1] But Delondre's poetic tribute must have broken the curse, at least temporarily, for both men returned safely—Delondre to his factory, and Weddell to France. In so doing, Weddell succeeded where La Condamine and de Jussieu had failed, for in his bags he carried the first cinchona seeds ever to be exported from South America. Some of those seeds germinated in the Jardin des Plantes in Paris, while others were given to the Horticultural Society in London, and one to the Dutch in Java. However, the spirits that protected the tree had simply been toying with Weddell, for subsequent assays on the bark showed that they contained very little quinine. What is more, when he went to Bolivia a second time his luck deserted him, and on this occasion Ledger was a witness.

BY 1851, LEDGER WAS becoming increasingly aware of the value not only of cinchona but also of alpacas. A miniature cousin of the camel, the alpaca was first domesticated by the Incas and grazes on the slopes of the Andean highlands. It is graceful and docile, and its wool, which has a beautiful sheen, is highly prized for both its warmth and its lightness. In 1836 the Bradford mills had overcome the problems of spinning alpaca wool, and high prices were being paid for the fleeces in Europe. A number of attempts had been made to introduce the animals into England and other countries, but few survived the sea voyage. In 1845, for instance, the *Sir Charles Napier* had left the coast of Peru with four hundred alpacas on board, but when it reached the Thames only three were still alive. To the Indians, for whom the alpaca was sacred, the news was a bad omen. "I shall never forget the commotion among the Indians on the news of this disaster," wrote Ledger. "Every misfortune that had happened in the district for the year past was attributed to these animals being sent out of the country."

Nevertheless, Ledger—who had given up on cinchona for the moment—was convinced that alpacas held the key to his fortune, and he began crossing them with llamas to increase their size. Next he began making inquiries about the climate and terrain in New South Wales, with a view to exporting the animals to Australia, from where, he calculated, he would be able to raise vast herds to supply the British Empire.

There was one problem: just as the Bolivians had set up a national bark bank to preserve their *calisaya* forests, so the Peruvians had banned the export of alpacas in a bid to protect their monopoly on the supply of wool. Penalties for those breaking the law were harsh—ten years hard labor on the guano islands.

In 1851 the Lima authorities turned away an Australian ship chartered for the express purpose of conveying a cargo of live alpacas to New South Wales. Ledger noted the incident with interest. He knew the Indians who bred the alpacas in the highlands better than any man. He also knew the country. Now, during his travels to the Bolivian frontier, he formulated an audacious plan. It was illegal to drive alpacas within forty miles of the coast of Peru, but there were no such constraints against their export from the coast of Chile or Brazil. The reason was that only a madman, or someone as determined as Ledger, would consider attempting such a feat. To reach Chile, the alpacas would have to be driven across the Bolivian border, south through Bolivia into Argentina, and then over the Chilean Andes to the Pacific coast. The only other possibility was to drive the alpacas over the eastern range of the Bolivian cordillera down to the Amazonian plain, from where he could transport them by river to the Atlantic.

Of the two, the second option was probably the less likely to succeed. Even if the alpacas survived the treacherous descent to the Beni River, there was no guarantee they would survive the heat and humidity of the jungle. However, this is the route Ledger and Mamani now decided to explore. Ironically, by taking this route they would answer a question that had so far eluded Weddell—namely, what was the variety of cinchona with the highest level of quinine?

IN 1851, WEDDELL HAD RETURNED to South America with the intention of entering the Bolivian cinchona region of Tipuani and gathering plants and seeds of *roja*. After descending the eastern slopes of the Andes, Weddell sailed down the Tipuani River to Guanay, a mission of Lecos Indians, and then ascended the Coroico in a canoe. The riverbanks were crowded with gray cinchona and *Cascarilla magnifolia*, an allied genus with deliciously fragrant flowers. But the bark hunters considered both varieties "*carhua-carhua*"—worthless. To find the true *calisaya*, they had to venture deep into the forest, an undertaking that Weddell now discovered was fraught with danger.

Weddell had landed on a beach on the Coroico with the intention of camping for the night when he chanced upon the hut of a *cascarillero*. He had taken only a few paces when he spotted a man stretched out on the ground. Nearly naked, and covered with insects, the *cascarillero* was in his death throes. He had been stung all over his body, and his face was so swollen that to Weddell it "no longer appeared human." On the roof of his hut were his only belongings: a straw hat, some ragged clothes, a knife, and an earthen pot containing a little maize and two or three *chunus* (dehydrated potato snacks)—his last supper.

The image of the dead *cascarillero* haunted Weddell all the way back to Peru. It must have been an omen. Certainly, that is the way it looked to Ledger when, for the first and only time, he met Weddell at Islay as he was about to embark for Europe. Ledger helped Weddell load two Wardian cases packed with *calisayas* he had collected in Bolivia onto the ship. But when they opened the cases to check the plants' condition, Weddell burst into tears. All the best specimens were dead. Only five small plants were still alive, but they were so sickly they stood little hope of surviving the voyage. "After five months of immense anxiety, hard work, personal dangers and great expenditure, I could well understand how a strong man could be so affected," wrote Ledger.

Weddell was too distraught to return to the forest. His mood was not helped by Ledger's description of a huge old *roja* tree that he and Mamani had seen growing at the house of an Italian curate in Apolobamba, in northern Bolivia. Feeling sorry for Weddell, Ledger agreed to collect some seeds from the tree during his next visit and forward them to him in France, but, incredibly, no sooner had Ledger gathered the seeds and given them to a friend in Arequipa for safe-keeping than the friend's house burned down. Even so simple and innocent a request, it seemed, was doomed to failure. Weddell had promised to return to Peru in 1853 and go with Ledger to Caupolicán, but after this incident his spirit was broken and, as Ledger writes, "I heard no more from him."

Ledger was not so easily discouraged. Moreover, he knew the Indians. He had previously noted how Mamani had been reluctant to collect seeds for export. Now, sitting around the campfire, he questioned him more closely. To protect their livelihood, Mamani explained, the *cascarilleros* would often appear to do the gringo's bidding while secretly sabotaging the operation. A favorite trick was to expose the seeds to heat so that they became sterile; another was to mix the soil in the Wardian boxes with arsenic so that the plants would die on the voyage home.[2] Sensing Mamani's inner conflict, Ledger did not pursue the subject, but the warning was noted. No wonder it was so difficult to smuggle cinchonas out of South America. Even if European botanists worked out which were the most valuable varieties and arrived at the right time to collect the seeds, the *cascarilleros* would be on hand to sabotage them.

WHETHER OUT OF SUPERSTITION or fear of discovery of his true intentions, Ledger now began to exercise extreme caution. In 1842 he had married Candelaria Ortiz, a woman from an influential local family in Tacna, and by the summer of 1851 she had borne him three daughters, Elena, Eliza, and Isidora. In addition, his younger brother, Arthur, had joined him from England. If Ledger's plans for the export of alpacas or cinchona became known, not only his life but also those of his family and even Mamani's could be in danger. Perhaps that is why Ledger's notes from this period suddenly become sketchier and less precise, and why, when he wrote to friends in London, such as John Howard, a prominent quinine manufacturer in Tottenham, he left blank the names of people and locations.

From piecing his letters and journals together, however, it appears that Ledger and Mamani set off to explore a Bolivian passage for alpacas in October or November of 1851. Starting from Puno on the west side of Lake Titicaca, they followed the same route as Weddell, through the province of Caupolicán and down to the Bolivian Amazon. Even for two such hardy travelers, it was an arduous journey. The banks of the Amazon's upper effluents were already dangerously swollen, and even small mountain streams had been transformed into violent watercourses. Ahead of them lay a succession of almost impassable steep rocks and forests tangled with roots and vines. The route was so overgrown it was impossible for Ledger's mule to find a way through, and soon he and Mamani were walking side by side, both of them chewing coca to keep up their stamina and to steady their heads in the dizzying altitude. After days of hacking through the forest, they were about to give up hope when they reached a point where the headwaters of the Beni and Mamoré rivers run close together. Suddenly, on a slope behind a bend of one of the rivers—from Ledger's notes it is not clear which—they stumbled on a group of fifty huge *roja* trees in full bloom. Usually *calisayas* were found intermixed with other types of cinchona around a bigger and older *tata* (father) tree, but these were all alike and clustered together. Brilliant-colored hummingbirds flitted in and out of the branches, savoring the delicate lilac-scented flowers. Combined with the scarlet of the leaves, the sight must have been breathtaking. Never had Mamani, let alone Ledger, seen such a magnificent group of cinchonas.

Because the trees were in bloom, it was too early to collect their seeds, and in view of the near-impassability of the terrain it was also impossible to harvest the bark. Besides, Ledger had the more pressing business of the alpacas to attend to. Nevertheless, he would never forget the image of those *calisayas*. The vision would stay with him for eleven long years—a period during which no one would lay eyes on cinchonas of their equal again.

Unfortunately, the exploratory route for alpacas proved a failure. The only other way to get the animals out of Peru was to avoid the

Bolivian cloud forest and take them across the altiplano to Argentina and then across the cordillera instead. But it was a long, difficult, and dangerous route—a last resort at best. The only other possibility was illegal: driving the alpacas from the highlands across a forbidden forty-mile strip of land directly to the Peruvian coast. Normally, this would have been out of the question, but it appears that Ledger received encouragement from an unexpected quarter: the British government. I say "appears" because the records are fragmentary and, according to Ledger, the impetus came from Mr. Pitt-Adams, Her Majesty's chargé d'affaires in Lima. Given that Britain and Peru were at peace, this was highly improper. According to Ledger, the reason was an Australian protest to British ministers in London about the recent expulsion of one of their vessels. However, there is no record of Pitt-Adams's having put anything in writing, and Ledger's claim of official involvement may simply have been a fabrication to cover his subsequent actions. If so, it would not be wholly out of character—Ledger's later career and writings indicate that he was given to grandiose ideas and sweeping statements.

At this point Ledger's diary becomes even more vague. All that is known is that in March 1853 he left Tacna for Chulluncayani, the estate where he had been raising alpacas, and began "putting in motion" part of his herd. Avoiding the road that runs between Tacna and the coast, Ledger rode up the western cordillera toward the border with Bolivia. Presumably he was still exploring possible routes, but here his plans must have been betrayed, or something else went wrong, for his diary records that he was detained at the frontier until September—in other words, for six months. It seems that during this period he ran into Hasskarl at Cojata. Despite the lengths he had taken to conceal his identity, Hasskarl dropped his pretense and, according to Ledger, confided the true purpose of his mission. Although Ledger was preoccupied with alpacas, that conversation rekindled his interest in cinchona. For Hasskarl confided that the Dutch had been willing not only to underwrite all his expenses, but also to provide him with a frigate for the return journey. Years later Ledger's memory of that conversation would become a source of bitterness—one more debit to be added to his tally against the Dutch and British.

The seeds of Ledger's resentment had been planted in 1845, after the failure of the expedition to Carabaya. Realizing that supplies of cinchona were in increasingly short supply, he had written to his father in London "urging him to solicit the attention of the East India Company or the Government of Great Britain" for a plan to raise *Cinchona calisaya* in India, Jamaica, and Ceylon. At the time, imports of quinine to India totaled some eighteen thousand pounds a year. Indicating how the plants and seed could be obtained, Ledger had pointed out that all it would take to cut Britain's supply line was "a miserable war between Peru and Bolivia."

Ledger did not even receive the courtesy of a reply from the British government. Marooned in Peru five thousand miles from home, with neither connections nor influence, he was a nobody, just another distant subject of the great British Empire. Perhaps that is why he latched onto the alpaca. The Australians were interested and, if he is to be believed, so were the British. But before gambling everything on such a risky venture, he had to be sure that conditions in New South Wales were right for breeding the animals.

He spent the next year at Tacna planning his trip to Australia. His mind was made up: the way to fortune lay in alpaca, not cinchona. Ironically, his obsession with the revered animal of the Incas would lead him to miss the arrival in Peru of an enthusiastic young Inca scholar, a man who would soon be in a position to provide Ledger with the very entrée into the British establishment he so desperately craved.

5

The Philanthropist

The almost inappreciable value of the cinchona, commonly called Peruvian bark . . . is universally acknowledged. Hence it becomes a duty to humanity . . . to increase the supply of the cinchona trees which yield such valuable barks."

—*Memorandum from Forbes Royle to East India House, March 1857*

AT THE HEADQUARTERS of the Royal Geographical Society in Kensington, London, hang portraits of its distinguished past presidents: Earl Curzon of Kedlestone, Lawrence Dundas, second Marquess of Zetland, and Sir Clements Markham. The paintings are for the most part dull and uninteresting—oils so darkly rendered they strain the eye. However, the one of Markham repays scrutiny.

Markham was an explorer of the old school. When he sat for the artist, George Henry, the year was 1913 and "Clemmie"—as he was known to his family and friends—was eighty-three. The society's longest-serving president ever, Markham had been geographer to the Abyssinia Expedition of 1868, and the guiding light behind Scott's two expeditions to the Antarctic. It is the last for which he is best remembered, his patronage earning him the reputation as the father of polar exploration.

But surprisingly, his portrait contains only one allusion to the poles—a small silver sleigh—and none to Scott's ship, the *Discovery*, which he both fitted out and brought back after it became trapped in the ice. Instead, on a wall beside the chair in which Markham is pictured, the artist has rendered a painting of *Cinchona uritusinga*. The species is credited to Pavon—the Spanish botanist who accompanied Ruiz on the tragic expedition to Peru and Chile in 1778. But it is by its original name, *Cinchona officinalis*, or more colloquially, "*cascarilla fina*," that it is better known.[1] It is, of course, the famous red bark tree of Loja, the one that supposedly cured the condesa and whose seeds—along with those of the *calisaya*—were deemed so valuable by the British and Dutch.

The reference is no accident. Although his role in the quest for cinchona is all but forgotten today, Markham considered it his finest hour. He may not have possessed Spruce's botanical expertise, or had Ledger's local knowledge of Peru and Bolivia, but he more than made up for it with his vision and will to succeed. More important, Markham had the right political connections. It was those connections that made him the focus—at least as far as the public was concerned—of the quest for quinine. Markham was both the lightning rod and facilitator of that hunt. His genius was to present the plot to steal the seeds and plants from under the noses of the nascent Andean republics not as theft but as philanthropy. To some extent his argument was justified; by the 1850s the South American bark forests were rapidly being stripped to exhaustion. More significantly, it was Markham who brought Spruce into the hunt. He could have advanced Ledger's cause, too, but though they shared similar aims, their paths never crossed. And by the time Markham learned of Ledger's expertise, it was too late.

BORN ON JULY 20, 1830, at Stillingfleet, in the West Riding of Yorkshire—no more than fifty miles from Spruce's hometown of Ganthorpe—Clements Robert Markham came from a long line of churchmen. His great-grandfather, William Markham, had been archbishop of York and tutor to both George IV and William IV. In 1827 the royal connection had brought Clements's father, the Reverend David Markham, the canonry at Windsor. The post required him to spend two months of the year at the castle, where he was expected to dine every Sunday with the king and queen. As a boy, therefore, Markham often found himself playing with more privileged children from Eton, an experience that no doubt gave him certain expectations in life.

His key role model, however, was his father. Tall, handsome, and a keen cricketer, the Reverend Markham was not your stereotypical country vicar. He excelled at carpentry and was a keen student of both medicine and numismatics, and it is from him that Markham seems to have acquired his love of academia. Educated at Cheam and later at Westminster, Markham was a precocious student. By the age of eleven he had written histories of England and all the ancient civilizations of the world, and a year later his interests had expanded to include historical biography, astronomy, geography, and natural history. But it was in the navy that Markham acquired his lifelong passion: exploration. And his entrée came about because of family connections.

In 1844, after only two years at Westminster, Markham was invited to dinner at the house of his aunt, the Countess of Mansfield, in Portland Place. Either by luck or design he found himself seated next to Rear Admiral Sir George Seymour, one of the Lords of the Admiralty. In the course of the conversation, the admiral innocently asked whether Markham would like to join the navy and come out with him to the Pacific in his flagship, HMS *Collingwood*. Not realizing the full import of the offer, Markham said yes. Thus it was that at the age of fourteen he found himself enrolled as a navy cadet and setting sail for Peru, via Madeira, Brazil, and the Falkland Islands. It would be the first of three visits to the land of the Incas, and on each occasion he would be drawn deeper and deeper into the country and its history.

Markham quickly proved himself an able cadet as well as something of a favorite with the officers. This, it seems, was due to both his looks and his personality. He had fine, delicate features and, according to one officer, was easily "the most engaging boy on board the ship." But Markham found it difficult to adjust to the harsh naval regime, with its strict rules and regulations. Raised in a spirit of Christian forgiveness, he sometimes found the insistence on strict discipline and punishment at sea cruel and barbaric. And when, in 1845, a lieutenant to whom he had become particularly close was suddenly transferred to another ship, he wrote that his heart was "like lead" and that he "shed bitter tears."

But for all his misgivings about the navy, Markham loved the sense of adventure that went with seafaring and wrote that his first sight of Lima had filled him with "delight." As soon as his naval duties allowed, he set off on a tour of the Peruvian capital, visiting the crypt where Pizarro was buried and immersing himself in the histories of the conquest. HMS *Collingwood* sailed on, but Markham left his heart in Peru, vowing to return one day to Cusco, the ancient Inca capital, and to write a history of its people.

Although he now wanted to leave the navy, his father was against it and Markham dared not go against his wishes. Besides, the navy still offered the consolation of travel, and in 1850 he found himself in HMS *Assistance* as part of an expedition to the Arctic in search of the English explorer Sir John Franklin.

Franklin had set out five years earlier in an attempt to find the Northwest Passage linking the Atlantic with the Pacific. He had not been heard from since. In fact, it was later discovered, Franklin and his crew had frozen to death in 1847–48. Markham would later draw on his experiences in pursuit of Franklin to further the cause of polar exploration, but for the moment it was not the poles but South America that he was obsessed with. Stuck in the Arctic ice during the winter of 1850–51, Markham seized the opportunity to reread W. H. Prescott's classic history of Peru until he knew it "almost by heart." He also began studying a Quechua grammar, which he had picked up in Lima. By the time he returned from the Arctic in September 1851, Markham's mind was made up. Telling his parents that "all my wishes, hopes and aspirations are entirely apart from the Navy," he resigned his commission. However, it was not until the following summer that he plucked up the courage to tell his father of his new ambition, namely "to go again to South America and search the interior for Peruvian remains." The reverend had set his sights on his son's studying the law and at first opposed the plan, fearing that "Clemmie" would not be able to support himself as a writer. Eventually, however, he relented, giving Markham £500 toward his expenses and using his influence to help him obtain letters of introduction to various dignitaries and officials in Peru. Thus it was that on August 20, Markham left Windsor for Liverpool, crossing the Atlantic with two former messmates from the *Collingwood*. He would never see his father again.

Markham did not head straight for South America. Instead, he crossed to Halifax and proceeded along the coast of North America to New Brunswick and Boston in order to call on Prescott, the Inca historian. They spent ten days together. Between hands of whist, Prescott unhesitatingly approved Markham's expedition to Cusco, the "navel" of the Inca empire, saying that unless he was personally acquainted with the places and people he sought to describe he could not expect to write an informed history. However, the delay meant Markham did not set sail from New York until September. It was the beginning of a ten-month visit that would take him not only to Cusco but as far east as the cinchona forests of Urubamba on the Madre de Dios River. Unlike Ledger, Markham at this time knew little of the history of cinchona. He was also wholly ignorant of the importance that the India Office was beginning to attach to the seeds and plants. Later, after he became a part of what he fondly called the great work, it would strike him as ironic that at the moment Britain was making its first efforts to obtain cinchona seeds, he had been riding through the very valleys where some of the best trees grew.

MARKHAM'S AIM IN VISITING PERU was to survey the seat of Inca civilization and to elucidate the Incas' history by interweaving it with a narrative of his own travels—a very "modern" ambition. This project would take him along Peru's narrow coastal strip as far as Nazca, site of the famous Nazca lines, over the Andes to Cusco, and down the eastern slopes of the cordilleras to the Peruvian Amazon. During his journey Markham was received by past, present, and future presidents of Peru and mixed with Peruvians of all classes and racial backgrounds. The trip resulted in two books, *Travels in Peru in 1853*, a two-volume account of his travels complete with drawings and watercolors, and *Cusco and Lima*, a pedagogical introduction to Peruvian history. More important, from the point of view of his later role in the hunt for cinchona, it gave him an insight into Peruvian politics and the highland Indians, many of them descendants of the Incas and a people for whom he felt a particular affinity.

Markham began his descent of the eastern Andes at the end of April, the beginning of the dry season, reaching Paucartambo, just above the Peruvian Amazon, on May 2. Intent on following an old Inca trail, Markham first climbed to thirteen thousand feet and then began a difficult zigzag descent to the rain forest. It was here that he saw his first *cascarilla* tree—*Cinchona ovata*, the same species Hasskarl would later import to Java with such disastrous results. Ignorant as he was of botany, even Markham realized that this was "not a valuable kind." However, it was only when he reached the Madre de Dios River, an upper tributary of the Beni and the same river on whose banks in Bolivia Ledger had spotted the rare group of *calisayas* the year before, that he had his first inkling of cinchona's potential.

Markham had gone to a hacienda in San Miguel to make the acquaintance of a Carmelite friar, an Italian by the name of Father de Revello. De Revello had spent six years in Paucartambo, and as well as being an authority on the Incas, was intimately acquainted with the *montaña*. Eight years previously a priest had been murdered at a farm overlooking the Madre de Dios—stuck through with arrows by wild Chucho Indians. The tale so alarmed Markham's guide that he refused to continue. But Markham was determined to see the river for himself and set off with Father de Revello alone. By the time they reached the farm where the priest had been murdered, it was noon and the sun was "scorching." To view the Madre de Dios, they had to cut through the jungle between the farm and the river with their machetes. It was worth the effort. The river, writes Markham, was a "grand sight," and in the evening when they returned to the farm, Father de Revello talked about his recently published pamphlet, *El brillante porvenir del Cusco*—"The Splendid Future of Cusco." "He spoke of cinchona trees yielding quinine . . . of the Indian rubber trees, a source of boundless riches, of the balsam trees, the chonta palms, and bamboos and their many uses . . . Then in imagination he called up before his mind's eye the thousands of square miles that might be cultivated with coca, maize, coffee, cocoa, bananas, sugarcane and cotton and the happy Christian population such crops would sustain in comfort." Father de

Revello's agricultural utopia would never be realized, of course—that dream had disappeared with the Incas. But long after Markham returned to England he would recall the "wealth of the *montaña*." Father de Revello had planted a seed in Markham's mind—of the potential not only of cinchona but of rubber. That seed would grow.

Markham returned to Cusco on May 16 and four days later reached the coast at Islay, where he boarded a steamer for Callao. In a few days, he reached Lima. It was there he learned that his father was dead. He heard the news in the cruelest way, through an obituary in the *Times*, and hurried back to England to be with his family. The homecoming brought Markham back to earth with a jolt. His father had been both a friend and a mentor. Although he had never stopped Markham from pursuing his bliss, he had always urged on him the importance of a proper vocation. Now that the exhilaration of the Andes was over, Markham had to face reality. He was broke. In order to have time to write his histories, he would have to find a paying job, and quickly.

Fate now conspired to give him one of the few jobs at which he could both be financially secure and realize his wider ambitions. At first glance the Legacy Duty Office of the Inland Revenue, where Markham was offered the post of junior clerk, looked like a dead end. The job required him to spend his days in the gloom and dust of Somerset House, filling out ponderous registers and preparing indexes. But because of a bureaucratic anachronism, the Inland Revenue was also responsible for the department of the Board of Control, an office that worked closely with the East India Company in administering the government of India. After just six months at Somerset House, Markham was offered the chance to transfer to the board's Secret and Confidential Branch. It was the perfect posting for him. Here Markham would be responsible for copying letters and dispatches from India, Persia, Syria, and the Orient, and still have free time to write his histories. More important, he would also come to the attention of the officials who ran the India Office.

BY THE 1850S as many as two million adults a year were dying of malaria in India, and a further twenty-five million had been afflicted with the disease at one time or another. Partly this was a result of the East India Company's disastrous "land tax" policy, which had mired the peasantry in debt and led to the neglect of roads, canals, and river works essential to controlling mosquito larvae. By the middle of the decade, the company was spending £100,000 a year on importing bark from South America for its employees, but it was not simply a question of cost, it was also a matter of supply. Writing to India House in 1852, Forbes Royle, the reporter on Indian products to the East India Company, had warned that rather than stabilizing the price of bark and protecting the country's monopoly, Bolivia's national bark bank was leading to rapid inflation and the destruction of the *calisaya* forests.[2]

At the same time, the British were also beginning to realize the value of quinine in furthering their wider colonial ambitions. Nowhere was this truer than in the "white man's grave" of West and central Africa. From the Scottish shipbuilder MacGregor Laird's disastrous 1832 expedition up the Niger—when all but nine of his forty-nine crew members died of malaria—to the Niger Expedition ten years later (forty-two deaths out of a hundred and forty-five), British ambitions of penetrating the continent's "dark" interior had been impeded by tropical disease.

Interestingly, the first person to apply the medical lessons of these expeditions was the missionary explorer David Livingstone. While practicing medicine in Botswana, he had obtained a copy of the medical report of the 1842 Niger Expedition and noted the high incidence of malaria and the failure to administer quinine early enough. When he set out with his family for Makolo country in 1850, he made sure he brought ample supplies of quinine—which almost certainly saved the lives of his son and daughter when they were suddenly stricken with malaria at Lake Ngami in northern Botswana. Two years later, traveling alone this time, Livingstone also contracted malaria and cured himself with quinine. The following year, in May 1853—a few months before Spruce contracted malaria in Venezuela—he suffered another bout of fever at Luanda. From then on he took quinine regularly, both as a curative and a prophylactic, enabling him to survive the malaria-infested Zambezi and numerous other expeditions to central and southern Africa over a further period of twenty years.[3]

Livingstone was not the only medic to laud the prophylactic benefits of quinine. Although the importance of administering quinine wine to sailors *before* they disembarked for work ashore had been recognized for some time, it gained new urgency from the West African squadron's experience on the Gold Coast and in the Bight of Benin, the notoriously malarious seas surrounding the vast Niger delta, which inspired the sailor's refrain "Beware and take care of the Bight of Benin/There's one comes out for forty goes in."

In 1854, William Balfour Baikie, a Scottish naval surgeon in charge of yet another expedition to the Niger, instituted a strict regime of quinine prophylaxis. His vessel, the *Pleiad*, was therefore able to make it to the confluence of the Niger and Benue rivers, from where Baikie was able to chart a further 240 miles of passable river to the east, and back again without a single loss of life (indeed, only two Europeans fell ill). Although Baikie—like all doctors at the time—remained ignorant of the connection between the mosquito and malaria, his conclusion was unequivocal: "In no essential does African endemic fever differ from the fever of Hindustan, of Borneo, of the Spanish Main, of the West Indies, or of fenny and marshy countries in Europe." Moreover, in each case Baikie believed there was only one treatment—quinine.

In 1848 the army's medical department had also taken the first tentative steps toward quinine prophylaxis, directing its governors in West Africa in the use of quinine to prevent fever. And by the time of the Crimean War five years later, quinine had become a standard precaution for both cavalry and infantry.[4]

As quinine's value became more and more obvious, Royle warmed to his theme. In March 1857 he wrote again to India House, arguing that it was now the government's "duty to humanity" to gather cinchona seeds before the forests were stripped bare and raise the plant in India. "The subject is yet of such great importance, both in an economical and philanthropic point of view, that every exertion should be made to ensure its accomplishment." It was a clever argument. Eager to maintain its reputation as the powerhouse of British botany, Kew Gardens had been campaigning for some time for the collection and cultivation of all the varieties of cinchona growing in the Andes. By recasting the theft as philanthropy, Royle had provided the bureaucrats at the India Office with the perfect justification for putting Kew's plan into action.

It was against this background that the Court of Directors of the East India Company approved Royle's proposal to send an experienced plant collector to the cinchona forests. But Royle died before his plan could be acted on. Then in May 1857 came the Indian Mutiny, plunging the British administration of India into crisis. Responsibility for the governance of India passed from the East India Company to the Crown, and Royle's scheme was momentarily forgotten in the reshuffle of responsibilities. In November 1858, Kew's director, Sir William Hooker, tried to inject new urgency into the plan, writing to the India Office to warn that "the Dutch have already got plantations of cinchona trees in Java 30 feet high, and we not a live plant in India for want of some energetic practical men to direct such operations." The implied criticism of the Foreign Office's earlier lackadaisical approach was clear. The task was now too important to be left to the South American consuls. But it was only with the appointment of Royle's successor, Forbes Watson, that in March of the following year Lord Stanley, the secretary of state for India, agreed to an expedition to South America for the sole purpose of obtaining cinchona. The question was, who was to lead it?

By now Markham had spent five years as a clerk in the Board of Control. He had used the time profitably, completing his history, *Cusco and Lima*, and contributing papers to the Royal Geographical Society, which established his reputation as a Peru expert and explorer. In the summer of 1856 he had also met the love of his life, Minna Chichester. The daughter of the Reverend James Chichester, rector of Arlington, Minna shared Markham's ecclesiastical upbringing and, like him, was a skilled linguist. They were married within a year and became lifelong companions, Minna accompanying Markham on many of his travels and helping him with his translations.

It is not clear when Markham first became aware of the importance of cinchona to India. It would be surprising if through his position in the Board of Control he was not familiar with the growing correspondence on the subject between the India Office and the home government. However, it was only when Henry Deedes, assistant secretary in the India Office's Public Department, approached him with the proposal that he should lead the expedition that Markham became an enthusiastic convert to Royle's arguments. "I gave the subject most careful consideration," Markham later recalled, "and being convinced that this measure would confer an inestimable benefit on British India, and on the world generally, I resolved to undertake its execution."

It wasn't quite that simple, of course. Markham had no botanical qualifications and first had to persuade the Council of India that he was the right man for the job. In April he began a determined letter-writing campaign, reminding the council's chairman of revenue and the undersecretary of state for India of his facility with Spanish and Quechua and his familiarity with the bark forests of Peru and Bolivia as well as with the landowners of the eastern cordilleras. He argued that failure in the past had resulted from leaving the work to agents and that great care and personal attention was needed during the collection and transportation of the seeds and plants to ensure they survived the journey and propagated. It was "necessary," wrote Markham, to select someone "whose heart is really in the business," adding that he was prepared to receive only his salary as a junior clerk (£250 a year plus expenses).

Not everyone was convinced of Markham's suitability for the task. The *Gardener's Chronicle* predicted that his efforts would fail because of his ignorance of botany and science in general. But Markham's enthusiasm overrode any doubts his superiors may have harbored about his lack of qualifications, and on April 8, Lord Stanley entrusted him with the commission.

It was a daunting responsibility for one so young. Markham was still just twenty-nine. Yes, he had visited one of the areas in Peru where cinchonas grew, but his familiarity with the trees was cursory. Unlike Ledger, he had never been to Bolivia, and he certainly could not tell the difference between the red bark from Ecuador and the more valuable *calisayas* from Carabaya and Apolobamba. More important, Markham was a novice in the art of deception. During his visit to study Inca remains, the Peruvians had gone out of their way to welcome him. This time, however, they would put every obstacle in his path. In short, it was the sort of commission that could make or break a man's career. Succeed and he would be elevated to the ranks of the great and the good. Fail and he would be consigned forever to the dustbin of botany.

If Markham was worried about the consequences at this stage, however, he kept his concerns to himself. Instead, he threw himself into the study of cinchona. In the archives of the Royal Geographical Society two of Markham's notebooks from this period have survived. Lovingly bound in leather and labeled "Cinchona Notebooks I and II," the pages are crisp and yellowing but the ink has hardly faded. Their contents attest to Markham's enthusiasm and thoroughness. There are excerpts from letters to Weddell, notes of his conversations with John Howard, the Tottenham quinine manufacturer, and extracts from the archives at Kew. Markham has noted every variety of cinchona then known to botanists, beginning with their Linnaean classification and followed by their commercial name, bark type, and location. There are sketches of the leaves, styles, and anthers, details of the climate and soils in which the cinchonas grow, and notes of the latitudes and elevations at which the best varieties are found. But most impressive of all is Markham's scholarship. He has read La Condamine in the original French, been to Madrid to examine for himself the writings of Ruiz and Pavon, translated Hasskarl's account of his expedition from the Dutch. And, of course, he has consulted Humboldt.

Although it is true that Markham lacked the "right" botanical qualifications, it is clear from his notebooks that he worked hard to correct that deficit. More important, he had a firm grasp of cinchona's medical and commercial significance and, as a Christian with a strong humanistic streak, was able to embrace the philanthropic argument seemingly without cant. Markham claimed that his aim was nothing less than the introduction of a cheap form of the cinchona febrifuge into every *pansari*'s shop in India at a cost of one farthing or less. "Thus," he wrote, "countless multitudes will be saved from death or grievous suffering."

With this goal in sight, Markham was determined to leave nothing to chance. The Ecuadorian government had recently granted four and a half million acres of land between Loja and Cuenca to a private company in an attempt to reduce its foreign debt. But when the company had tried to interest the French and British in an exploratory mission aimed at transporting cinchonas from Ecuador to the East Indies and

Algiers, Forbes Watson had rejected the plan, arguing that the best cinchonas grew in Bolivia. Markham now boldly insisted that if his mission was to succeed, the collecting should not be confined simply to Bolivia but should extend throughout the whole cinchona range, as far north as Colombia. He proposed that rather than sending just himself, the India Office fund four separate but coordinated expeditions aimed at gathering the eight most valuable varieties of cinchona growing in the Andes. These included both the Bolivian *calisayas* and two varieties of red bark in Ecuador. In addition, collectors should be sent to Huánuco, the area in Peru that Ledger had explored unsuccessfully in 1845, for the so-called gray barks and to Popayán for the Colombian barks. "If the thing is worth doing at all, it is worth doing well," he wrote to the India Office in July.

> It is now merely intended to send me to Peru, with a gardening assistant, to procure plants and seeds, and to convey them to India in the best way I can. The dangers, accidents and hindrances which might thwart me at every turn decrease the chances of success, because the whole thing would depend entirely upon the luck of one individual. In my opinion four separate expeditions should be sent in four different directions, their collections should be received on board one vessel, and taken direct to India; and thus all eight more valuable species of cinchona would be obtained.

In the end Markham received permission for only three expeditions: his own, another to Huánuco, and a third to Ecuador. His aim was to penetrate deep into Bolivia and Carabaya in Peru in search of *calisayas*, while his "fellow-labourers" mopped up the other cinchona species, much as assistant gardeners might rake up leaves. But Markham misread the political situation in Bolivia, and had it not been for the success of the Ecuadorian expedition, there would be little reason for associating his name with cinchona. That there is, is due almost entirely to the labors of one man: Richard Spruce.

AFTER HIS NEAR DEATH EXPERIENCE with malaria at Maipures, Spruce had waved goodbye to the insect plagues on the Orinoco and returned to San Carlos. But although he was now back on the mosquito-free waters of the Río Negro, he faced new threats.

In November 1854, Spruce began working his way downriver to São Gabriel with the intention of continuing his botanizing in Brazil. But on the journey to the Amazon his Indian guides hatched a plot to steal his money and murder him, reasoning that no one would be surprised to hear that a man as ill as Spruce had died. Fortunately, Spruce overheard their conversation and foiled the plot by lying awake all night in his canoe with one arm around his money chest and a double-barreled shotgun propped on his knee. Reaching São Gabriel safely, he then descended the Amazon to Manaus and Santarém, where he caught a steamer back the way he had come to Peru. By the summer of 1855 he had established a collecting station at Tarapoto on the Marañón River and was busy acquainting himself with the local flora, including cinchona. It was his first chance to examine in nature the tree that had saved his life, and he seized every opportunity to do so.[5]

By now George Bentham and Sir William Hooker would have been aware of the importance that the government was beginning to attach to cinchona in India. As director of Kew, Hooker may not have had direct access to the India Office's reports, but through his many government contacts he would have known of its growing interest in the cinchona trade and would have been eager to find a role for Kew in the developing enterprise. Who better than Spruce, a qualified botanical collector familiar with Spanish and the Indian languages and already in the Peruvian Amazon, to volunteer for the task? For whatever reason, however, his name was not put forward. Perhaps Hooker's influence did not extend as far as he would have liked, or perhaps he and Bentham feared for their friend's health. By 1856, Spruce had already endured six hard years in the jungle. As he would later bitterly recall, much of the time had been spent scrunched up in a dugout ankle-deep in water, his head and body exposed to the elements. His diet was monotonous—manioc, yams, smoked fish, and sometimes, if he was lucky, a little beef or pork. He often had to do without sleep because of the heat, the attacks of venomous insects, or, what was even more dangerous, the threats posed by his fellow man. And as he now reminded Hooker: "The Venezuelans say very expressively that however completely a man may think he has thrown off the ague, there always remains a bit of it sticking to his backbone."

There may also have been more practical reasons for Kew's hesitation. Although as the crow flies Chimborazo is only five hundred miles from Tarapoto, it is a far from easy journey. Today it is possible to drive from Tarapoto to Jaén and then proceed north by road along the spine of the Andes to Ecuador's avenue of volcanoes. In 1857, however, Peru was in the throes of a bitter civil war and travel by road was impossible. This meant that the only way to reach Ecuador was by river. From Tarapoto, Spruce would have to descend the Marañón to the Huallaga and then ascend the Pastaza to the Bobonaza, before proceeding by foot to the cinchona forests on the slopes of the Ecuadorian Andes. This would entail fighting his way up a series of rapids and whirlpools, fording rivers swollen by rain, and cutting his way through dense jungle inhabited by head-hunting Jivaro Indians. In short, the terrain was so inhospitable that it would have been far simpler for him to return to Brazil via the Amazon and then take a ship around Cape Horn to Guayaquil. But in that case why not simply send a fully fit and healthy collector by steamer from England?

Spruce, however, was determined to reach Ecuador, if for no other reason than that he was broke. Although he had more than £100 on account with Singlehurst and Company, which processed his botanical collections, the money had yet to arrive, and he was tired of living on a shoestring. In December 1855 he wrote to Bentham from Tarapoto that he had met a traveler from Quito who had already ascended the Pastaza several times and was planning to return by the same route the following autumn. Recognizing that he would be risking "a painful and dangerous voyage," Spruce told Bentham that he was inclined to accompany him. Quito, he had heard, was "the most splendid station for a botanist in South America." Besides, after so many years on the Amazon, scrabbling for food and botanical paper, it would be "such a fine thing to have headquarters where every necessary care [can] be promptly obtained."

Interestingly, the only mention of cinchona at this stage came not from Bentham or Hooker but from a London druggist, Daniel Hanbury. The Hanburys had long had an interest in medical botany, establishing a physic garden on the Italian Riviera, and together with their nephews, the Allens, they ran a successful pharmaceutical manufacturing business in Spitalfields. On hearing of Spruce's arrival in Tarapoto, Hanbury had asked Bentham to pass on a letter introducing himself as a potential client. It was the beginning of a lifelong friendship. Between 1856 and 1875, Spruce and Hanbury exchanged more than five hundred letters, and toward the end of his life Spruce used the correspondence to give vent to what he saw as his mistreatment by the India Office and his bitterness at his poverty and ill health.

On May 10, 1856, Hanbury prepared a memorandum, reminding himself to draw Spruce's attention to cinchona, "a group of plants, whose importance to medicine, we are all aware." The memorandum urges Spruce not only to collect specimens of young cinchona, but of the "old bark rich in alkaloids," which owing to an "absurd prejudice" was being all but ignored by European traders. "Botanical specimens of Cinchona & its allies, accompanied by small, authenticated specimens of bark, could not fail to meet with some demand here and on the continent in quarters where the general herbarium would not be subscribed . . ."

Unfortunately, his letter to Spruce has been lost, but it must have contained similar wording. In enclosing Hanbury's letter of introduction to Spruce, Bentham writes that Hanbury "is a young man of great abilities who has devoted himself much to pharmacology and is anxious on the question of cinchonas etc upon which he writes to you." And in a follow-up letter in July, Bentham adds: "I am very desirous to do anything to assist him—and he is very ready to give any fair remuneration."

What effect, if any, this had on Spruce's thinking is unclear. As far as Spruce was concerned, Hanbury was just one more collector. Besides, he had no reason at this stage to believe that cinchona was going to prove more lucrative than any other plant. His motives were purely practical. He had money in the bank, but because of the war in Peru it was impossible to get his hands on it. In Ecuador, however, Bentham had arranged for him to pick up a draft from the local Lloyd's agent. If he could only get to Quito, he would have the freedom to work on his collections in a climate agreeable to his health. So on March 23, 1857, Spruce set off from Tarapoto on an epic three-month journey to

Baños in Ecuador via the Huallaga and Pastaza rivers. Teaming up with two Ecuadorian businessmen, Spruce hired twelve Indians, five for the Ecuadorians and seven to crew his forty-foot mahogany dugout. It was a journey that no one had attempted before and no one has attempted since. It was so arduous that when Spruce reached Baños three months later, he could "hardly bear to think of it."

The voyage began, as it would continue, with a rainstorm. The first stage, to Chasuta, was overland. In the dry season such a trek takes no more than a day, but it was so wet that it took Spruce and his party two. The storms had deepened the potholes in their path, and the track was so muddy that Spruce was obliged to walk barefoot to preserve his shoes. The result was that by the time he reached Chasuta he was both lame and suffering from fever.

At Chasuta, Spruce provisioned canoes for the descent of the Huallaga. The boats had to be open so as to be able to pass safely under the many waterfalls en route, but no sooner had the party begun descending the river than they faced a far more serious danger: whirlpools. It was all Spruce's crew could do to keep the canoe from capsizing. Unfortunately Spruce had his dog Sultan, whom he had reared from a pup, with him in the boat. According to Spruce, "the horrid roar of the waters" drove Sultan mad, and from then on he would neither drink nor eat. For six days Spruce watched over him in the hope that he would recover. But when Sultan, almost worn to a skeleton, began snapping at people and bolted for the woods "uttering the most unearthly sounds," Spruce wrote, "I saw all hope of saving him was vain, and was obliged to shoot him."

Sultan's death seemed to drain the life from Spruce. Sultan had accompanied him all throughout the Amazon. Now he was alone with only his Indian porters and the Ecuadorians for company. To make matters worse, the Huallaga was infested with *zancudos*, and his skin was soon pockmarked with bites and ulcerations from the constant scratching.

The next few days passed in a monochrome blur: the river was flat and monotonous and devoid of human habitation. Spruce found the journey "heart-sickening." Eventually they reached the mission at La Laguna, near the mouth of the Marañón, where La Condamine had stopped on his way down the Amazon a century before. Although it was two weeks before Easter, Spruce's porters were intent on performing their penitence for Holy Week before journeying further upriver. Accordingly, while Spruce bought fresh provisions, the Indians prepared whips dipped in pitch and embedded with bits of broken glass. On the Saturday before they were due to embark, the "penitents" assembled in the main street. Spruce writes he had never seen a more "horrible sight." The Indians applied the straps "unceasingly," flagellating themselves until their backs were "one mass of gore" in the belief that it would ensure their safe return from the perilous journey ahead. Noticing that they were drunk, Spruce doubted the value of their penitence. However, at noontime on Monday each man was seated at his oars, ready to row, without showing any outward sign of discomfort.

It is only twenty-five miles from the Marañón to the mouth of the Pastaza, but the river was so low, and food so scarce, it took Spruce five days to cover the distance. On April 9, three days out from La Laguna, he reached a small village, deserted save for a dozen cane houses and several sacks of rice. It turned out that the men had been murdered by wild forest Indians while cutting cane, and their wives and children had had to flee for their lives. From then on, the Indians refused to sleep on shore, and Spruce never left the canoe without his gun.

Once they had entered the Pastaza, it took a further fortnight to reach Andoas. The village, on the border of Ecuador, lay two hundred miles upriver. The Pastaza was practically deserted, and the only settlement they passed was infected with leprosy. Although it rained continually, the water was low and at several points the porters had to leap out of the canoes and drag them over the shallow beds by hand.

On April 29, Spruce finally reached Andoas. It took him five days to lay in fresh provisions and engage a new crew to take him up the Bobonaza River to Canelos. Then, on May 5, he was off again. With its narrow, winding course, the Bobonaza reminded him of the Casiquiare. The water was muddy and sluggish, and once again progress was slow. It was only when they reached the village of Sarayuca that the vegetation began to change and the scene became more picturesque: waterfalls and cliffs clad with ferns and mosses and "a Helicomia [*sic*] with . . . pendent scarlet and yellowish spikes."

Now the going became tougher. The river was swollen by melted snow cascading off the distant volcanic peaks of the Ecuadorian Andes, and the Indian rowers had to paddle against a six-knot current. On May 21 they reached Puca-yacu, a village perched 250 feet above the river. They tied the canoes to lianas overhanging the narrow sandbar, then climbed up to pay their respects to the governor. But when they returned to the boats at sundown, the sky began to darken ominously. At 9 p.m. the heavens opened. The storm caused the level of the river to rise rapidly, swamping the beach "in a roaring surge which broke under the canoes in whirlpools." Fearing the boats would break their moorings, the Indians leaped into the trees to secure the lianas higher up. But the surging waters now threatened to dash the boats against one another and it was all the Indians could do to hang on in the hope of averting disaster. "Assuredly I had slight hopes of living to see the day," writes Spruce, "and I shall for ever feel grateful to those Indians who, without any orders from us, stood through all the rain and storm of that fearful night, relaxing not a moment in their efforts to save our canoes from being carried away by the flood."

It was only at ten o'clock the following morning that the waters abated sufficiently for them to unload the boats and haul the sodden cargo up the cliff face. "Wearied to death," Spruce spent that night in the village. But the following day he was well enough to explore the plateau above the river. At his feet stretched the valley of the Bobonaza. Looking north and west was the deep gorge of the Pastaza and rolling away from it, at sundown, "the lofty cordillera . . . in cloudless majesty." Looking from south to north he could just make out the volcanoes Sangay and Altar and the bluff peak of Tungurahua. And there, "peeping over its left shoulder," was Chimborazo, "the monarch of the Andes."

Spruce was so taken with Puca-yacu he stayed three weeks, gathering mosses in the nearby woods and making a collection of the local beetles. Then, on June 10, to the accompaniment of the distant rumble of Sangay, he set off on the final leg of his epic journey.

Spruce reached Canelos in two days, but from here on, the gradients of the Bobonaza valley became steadily steeper and more dangerous. The only way to proceed was on foot, and to have any chance of reaching Baños he had to lighten the loads as much as possible. Accordingly, he entrusted his drugs, barometer, and other valuables to the Ecuadorians and packed his remaining belongings, including some precious drying papers, into five small boxes. Henceforth, he would have to confine himself to pressing mosses and lichens.

Spruce set off early on the morning of June 14, accompanied by eight Indian porters and the Ecuadorians. There were no paths, just slippery ascents leading up and over the ravines of the Bobonaza valley. To reach Baños, he would have to ford a series of streams separating the Bobonaza and the Pastaza, each one more swollen and treacherous than the last. The rain and mud were constant companions, and only a pair of rubber boots purchased from a German traveler in Canelos kept his feet dry. From the outset, his porters complained bitterly and insisted on frequent rest stops. Passing through Jivaro country, his annoyance at the delays was lessened, however, by the discovery that he was now in the mossiest place on earth. "Even the topmost twigs and the very branches were shaggy with mosses," he enthused. "Whenever rains, swollen streams, and grumbling Indians combined to overwhelm me with chagrin, I found reason to thank heaven which had enabled me to forget for the moment all my troubles in the contemplation of a simple moss."

The Indians, however, did not share his enthusiasm, and by June 20, they were fed up with struggling under the weight of the loads. Spruce had been forced to pay them in advance, and fearing they would mutiny and desert him in the middle of the night, he now had no choice but to get rid of all the drying papers that had not yet been used for pressing.

The next few days were sunny for once, and soon they came out on a plateau above the Pastaza. A series of steep ravines intersected by rushing mountain torrents now lay in their path. Using bamboo poles as bridges, they inched their way across each impediment until, finally, they reached the largest one of all—the river Topo. Here, Spruce writes, his worst fears were realized. It had been raining steadily for two days, and for as far as he could see the Topo "was one mass of foam." The thunder from the water was so great it "made the very ground shake to some distance from the bank." They were less than two hundred yards from the junction with the Pastaza, but there seemed no way across. The rocks that provided the usual support for the bamboo poles had been washed away. The only possibility was a group of three rocks higher upstream, but these too were now partially submerged beneath the surging waves. The only choice was to wait and hope that the river fell, but after two days there had been little change. Spruce now faced a stark choice: send his porters back to Canelos for fresh provisions and risk their deserting, or risk a crossing. He called a council by the riverside and put it to the men that if they were game, there was a possibility of creating a bridge at the new group of rocks. The only problem was that because of the distance and the difference in height between the rocks, the bridge would necessarily slope and bend considerably in the middle.

The Indians agreed to give it a shot. Four bridges would be needed in all. The bridge from the bank to the first stone was short and completed with relative ease, but the second strained their ingenuity to the limit. Normally, bamboo would be levered from one rock to another and, once it was in place, a man would crawl along it dragging another bamboo secured with a liana to lie alongside it. Once four bamboos had been laid in this way, they could be lashed together and the construction of the next bridge could begin. In the case of the second bridge, however, the bamboos were too short and only reached the side of the rock. To add to the difficulties, the roar of the water was so deafening that Spruce could communicate only by means of hand signals. Eventually, the Indians succeeded in laying four bamboos, but by the time they had completed the third and fourth bridges, it was noon and the water level had started to rise again. If they were to cross they would have to be quick, but no sooner did they venture out onto the second bridge than it began to bend under the weight of Spruce's heavy boxes. Spruce had no choice but to abandon them by the riverside, trusting that he could send someone from Baños to retrieve them later. Even so, he barely made it. As he inched across the bamboos, a wave wet him "to the knees," washing away the bridge soon after.

Spruce's ordeal was almost over. The next day they waded for nearly a mile through "fetid mud" before descending to the shores of the Pastaza via a wooden ladder attached to the sheer cliff face. From here on it was simply a matter of following the course of the Pastaza north until they reached the falls of Agoyan. Normally, the falls are crossed by means of a stout bridge suspended forty feet above the river, but the rains in the sierra had been so great that the bridge had been washed away. All that remained were three or four trunks, of which only one was unbroken. Gingerly, they inched their way safely across. A few hours later they arrived at a farm and Spruce procured a horse. It was now just a short fifteen-mile trot to Baños.

Spruce reached the pretty spa town nestled beneath the Tungurahua volcano the same afternoon. It was July 1. In all, the journey from Tarapoto had taken a grueling 102 days. By now, Spruce's features lay shrunken behind a three-month-old beard. He was emaciated from lack of food and desperately needed to recuperate. But Baños was not what he expected. Located nearly six thousand feet above sea level, the air was so cold it immediately set his teeth chattering. Then came the catarrh, "with a cough so violent as often to bring up blood from both nose and mouth." He was disappointed to discover there was no drying paper to be had in the town, and, in addition, the house he was renting was on the same site as one that had been destroyed by an earthquake seven years previously. Nevertheless, he spent the next six months collecting ferns and mosses on the wooded slopes below Tungurahua, and in October he made a return trip to the Topo River and the forest of Canelos. (Earlier he had sent his porters back to the Topo to recover his belongings; they found them crawling with maggots but otherwise undisturbed.)

Ecuador, like Peru, was now in a state of political turmoil. War between the neighboring republics was imminent, and the Ecuadorian army was roaming the countryside pressing men, money, and horses into its ranks. Despite these difficulties, at the beginning of 1858, Spruce moved to Ambato, a town strategically located on the high sierra midway between Quito and Guayaquil. This was to be his collecting station for the next two and a half years. Soon afterward, he made an important acquaintance: James Taylor, who had been medical attendant to General Flores, Ecuador's former president. Born in Cumberland, Taylor had married the daughter of one of Bolívar's generals and was well placed to help Spruce. His relationship with Flores would prove particularly useful. Flores was a substantial landowner and one of his estates, on the western slopes of Chimborazo, was in the heart of the red bark region. For the moment, however, Spruce and Taylor confined themselves to discussing ancient Greek, Spruce coming away with the opinion that Taylor was "a very kind-hearted, honourable man, which can't be said of many Englishmen I have met in South America." Through Taylor, Spruce also met the American ambassador to Ecuador, Philo White. Whether Taylor and White also introduced Spruce to Walter Cope, the British consul in Quito who had gathered cinchona plants for the Foreign Office in 1852, Spruce does not say. But it seems that it was following a conversation with Cope that Spruce finally realized the importance to the British government of procuring cinchona.

Spruce stayed a month with Taylor, and then headed for Riobamba, where he visited a spectacular waterfall on the southern side of Tungurahua. However, the effort of wading up to the waist in the icy snowmelt under a hot sun took a predictable toll. Finding himself confined to bed for four days with fever and "rheumatic pains from head to foot," Spruce wrote Bentham a long, self-pitying letter. It is worth quoting in full, if only to underline his straitened circumstances, his continued ignorance of the cinchona effort, and the depth of his hypochondriacal self-doubt—somewhat ironic given the Herculean task he was about to undertake on behalf of Her Majesty's government.

> I should be very glad to return to England as you recommend me, to distribute my mosses, but I am fearful of again falling into delicate health if I go there. I have, besides, no funds beyond what are in your hands; these would soon be exhausted, and poverty is such a positive crime in England, that to be there without either money or lucrative employment is a contingency not to be reflected on without dread. On the other hand, I already feel myself unequal to the painful mountain ascents, exposed at the same time to a burning sun and a piercingly cold wind. The eastern slopes of the Andes no doubt contain much fine ground, but for want of roads they can scarcely be explored, except by one to whom the pecuniary value of his collections would be no object, and who could go to any amount of expense. I have often wished I could get some consular appointment here, were it only for £150 a year; but I have no powerful friends, without which a familiarity with the country, the inhabitants, and the languages go for little.

"Poverty is such a positive crime in England . . ." *Plus ça change.* In spite of Bentham's urgings, and in view of his "delicate health," it is not hard to see why he would have preferred the eternal spring of Ambato to the anonymity of damp, rain-swept Yorkshire, and why, in probing for some means of financial security, he would have sought a meeting with Cope in Quito. Not long after this meeting, in September 1858, Spruce wrote what can only be described as a begging letter to Cope's superior, George Lenox-Conyngham, chief clerk to the Foreign Office. A transcription of the letter is buried in Markham's Cinchona Notebook I. Markham never alludes to it in his published works on cinchona or in his later correspondence with the India Office on Spruce's behalf. Perhaps he thought it was too embarrassing—an individual of Spruce's qualifications touting for work—or perhaps it interfered with the image he wanted to project of himself as the architect of the quest. In the letter, Spruce writes that Cope informed him that "for want of some trustworthy person" familiar with cinchona, the Foreign Office had been unable to procure the specimens requested. Should the government so wish, then Spruce could "undertake to supply them" and "speedily transport myself" to the cinchona forests. Echoing Royle's arguments, Spruce continues: ". . . the bark trees are rapidly verging to extinction, and it is therefore of the first importance that steps should be taken for their cultivation on a large scale, in some country where there is no lack of industrious hands. I will therefore venture further to observe that should HM government think proper to employ me in this matter, they will have the advantage of a person already on the spot, and familiar with the country, the people and the language and (what is of more importance) with the plants whose products are sought."

Spruce's letter reached Lenox-Conyngham in December. It must have triggered an immediate enquiry to Kew along the lines of "Who the hell is Spruce?" for in January, Hooker wrote directly to the Foreign Office in support of his wayward botanist: "I will venture to say without hesitation if the Government ever makes an attempt of the kind there is no man so competent to the work as Mr Spruce."

But the Foreign Office was not calling the shots, the India Office was, and it was only with Markham's appointment in April that the wheels were set in motion. By now, the Ecuador Land Company, in its desperation to interest the British in its cinchona forests, was claiming that it had recruited Spruce. Markham quickly established that this was not true, and in May he was authorized to open negotiations directly with the Yorkshireman. Used for so long to botanizing on a shoestring, Spruce priced himself far too cheaply, asking only £30 a month—£10 more than he normally received from his subscribers. Unsurprisingly, the India Office leaped at the proposal. When the plant collector Robert Fortune had brought tea plants from China, he had been paid £500 for a year's work. Although transporting cinchona would cost the India Office greater outlays, given the hardships involved, Spruce's offer was a bargain. Later, this would become a source of some resentment to him; as usual, however, he was more preoccupied with his present health than his future livelihood. "I have been entrusted by the India Government with the charge of obtaining seeds and young plants of the different sorts of cinchona found in the Quitonian Andes," Spruce wrote to his colleague John Teasdale in April 1859. "This task will occupy me (if my life be spared) the greater part of next year."

Spruce was right to be apprehensive. Because of the huge demand for cinchona, the more accessible forests had already been stripped bare. To find virgin trees he would have to ride along the inter-Andean valley and then descend either the eastern or western cordillera and penetrate deep into the cloud forest. In the case of the eastern cordillera—where the climate alternates between spring and summer—this was relatively pleasant. But in the case of the western foothills of the Andes, where the Pacific weather system rules, the forests were subject to almost continual rain. The best time to approach them was between June and September—the South American winter, when the rains eased off. But these were also the coldest months on the Paramo de Puenevata, the plain on the northern side of Chimborazo, whose shoulder Spruce would have to skirt near the snow limit before dropping down to four thousand feet. The difficulties involved in such an expedition, wrote Spruce, "were not slight ones." Nevertheless, he decided that if he was to succeed, he needed to acquaint himself as soon as possible with the different varieties of bark.

So, with Ambato as his base, he began making a series of excursions, first to Riobamba, and then to Alausí south of the volcano Sangay, running the gauntlet of the army and deserting troops on the way.

MEANWHILE, BACK IN LONDON, Markham was making preparations for his own expedition. One of his first stops was the home of the renowned English quinologist John Eliot Howard. A staunch Quaker, Howard had been interested in cinchona ever since apprenticing in the family pharmaceutical business at the age of twenty. Quinine was fast becoming one of the firm's most important manufactures and, encouraged by his father, Howard began cultivating cinchona seeds in a greenhouse next to his Tottenham quinine works. In 1857, in recognition of his contribution to the study of the genus, Howard was made an honorary fellow of the Linnaean Society. The following year he purchased the original manuscripts of the Spanish botanist Pavon, together with the collection of cinchona specimens made during his ill-fated trip to Peru and Chile, and began a microscopic study of the barks. These studies would eventually lead to the publication in 1862 of his *Nueva Quinologia*—the most complete classification of cinchona at that time.

It is not known what Markham and Howard discussed, but there was no better authority on cinchona, and it is safe to assume that Markham questioned him closely on how to distinguish the species with the highest quinine content—namely, the red bark of Ecuador and the *calisayas* of Peru and Bolivia.

Next, Markham visited the Horticultural Society in Chiswick to consult Robert Fortune on the design of several sets of Wardian cases for conveying the cinchona plants to India. These would be brought out separately by expert gardeners sent from Kew to rendezvous with Spruce and Markham on the coasts of, respectively, Ecuador and Peru, once their collections were ready. Markham also lobbied for an Admiralty vessel to convey the cases speedily to India, or, failing that, one of the India government's own steamships. But no one would give him a firm commitment. In contrast to the Dutch, who had provided a frigate in an effort to ensure the success of Hasskarl's mission, the British attitude was almost cavalier. When Markham wrote to the

British consul in Lima requesting letters of introduction to the Peruvian bark companies, for instance, he was blithely informed, "I do not see any particular difficulties you would have to face in the prosecution of your mission." Markham was not nearly so confident. War between Peru and Bolivia was imminent. Moreover, he knew the *cascarilleros* who worked the highland forests were notoriously protective of their livelihood. Even if he reached the Bolivian border and succeeded in finding the "true" *calisaya*, it was by no means certain that he would be allowed to return with the seeds and plants.

Spruce's mood was also darkening. In August, after a two-week trek during which he was alternately frozen, scorched, and drenched, he had finally reached Lucma, a town deep in the bark forests of the western Andes, and seen the *cascarilla roja* for himself. His guide, Bermeo, had discovered a slender new shoot growing from an old felled tree. When Bermeo slit it with his cutlass, Spruce marveled as the colorless sap turned bright red on contact with the air. "The more rapidly this change is effected, and the deeper is the ultimate tinge assumed, the more precious is the bark presumed to be," he wrote. Because of the civil unrest, Bermeo was convinced that no one had been in the woods for months and that they would find better examples deeper in the forest. But Spruce did not think it was worth riding farther, and by October he was back in Ambato. Three months earlier he had been forced to hang out the Union Jack after government troops, their ranks swollen by black and mixed-race conscripts, had marched on the town following their victory over rebel forces. On that occasion the troops had "respected" his property, but, as he wrote to Hooker, the country was now in a "very unsettled state." All around, men and livestock were being pressed into the rival armies. To avoid being torn apart, many families had fled to the highlands with all their possessions strapped to mules. "The war, if it actually comes, will be something like what you have read in India, yet nobody knows what it is to be about," wrote Spruce, referring to the Indian Mutiny. "These Spanish republics are not unlikely to squabble among one another until—like the Kilkenny cats—there is nothing left of them but their tails, and then Jonathan will step in and make an easy prey of their mangled carcasses . . ."

Spruce may not have been thinking about cinchona when he wrote those words, but they would prove prescient. From Britain's point of view, the political instability in Peru and Ecuador presented her with the perfect opportunity to seize the seeds and plants from under the noses of the South American republics. And in this case, Spruce and Markham would be the Jonathans.

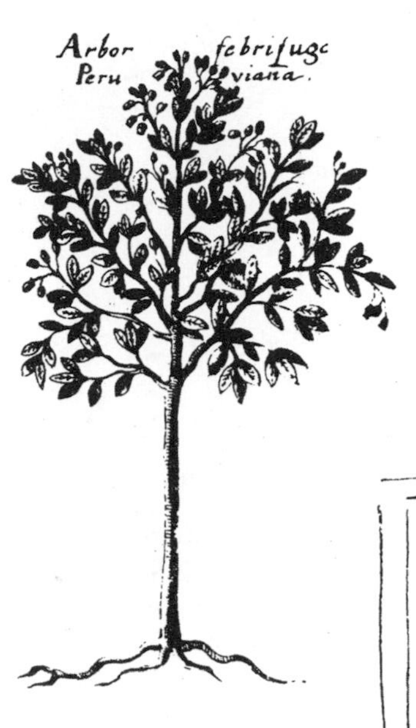

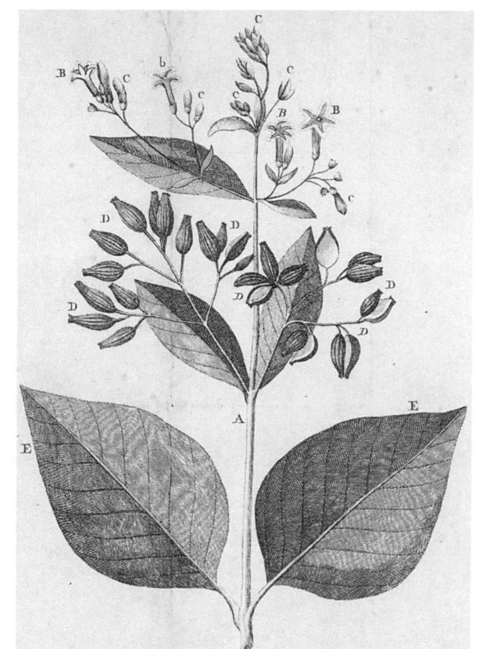

Above, left: One of the earliest illustrations of the "Peruvian tree of fevers," from Jonston's *Dendrographias*, published in 1662.

Above, right: Charles-Marie de La Condamine's sketch of *Cinchona officinalis*, from a 1745 Amsterdam edition of *Sur l'arbre du quinquina*. (Wellcome Library, London)

Right: By 1801, when this watercolor was painted, *Chinchona officinalis* was known as "the true Jesuits' bark." (Wellcome Library, London)

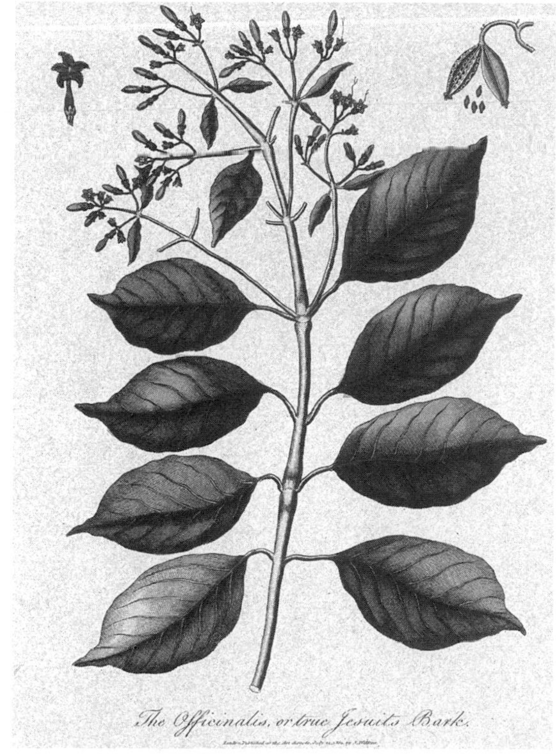

Oil-on-canvas triptych, after frescoes in the Santo Spirito hospital in Rome. Titled "Episodes in the History of Cinchona," the panels recount the legend of the Condesa's cure and show (*top*) the Conde de Chinchón instructing an Indian servant to collect cinchona bark for the royal pharmacy; (*middle*) the presentation of the "miraculous bark" to the Condesa; and (*bottom*) Cardinal de Lugo distributing the bark in Rome.

(Wellcome Library, London)

A woman gripped by the rigors of a malarial chill warms herself by an open hearth while the "ague monster" creeps up behind her in this 1788 colored etching by Dunthorne and Rowlandson. (Wellcome Library, London)

Richard Spruce, shortly after his return from South America in 1864.

Clements Markham, date unknown.

Charles Ledger, circa 1880
(Wellcome Library, London)

Manuel Incra Mamani, date unknown.

José Celestino Mutis, examining a flower petal in this 1930 copy of a nineteenth-century engraving. (Wellcome Library, London)

Alexander von Humboldt, pressing flowers in an 1806 portrait. (Wellcome Library, London)

Chimborazo, "the monarch of the Andes." (Copyright © Mark Honigsbaum)

The author's packhorses being loaded on the sierra prior to his trek over Chimborazo. (Copyright © Mark Honigsbaum)

A young *Cinchona succirubra* tree growing beneath Chimborazo.
(Copyright © Mark Honigsbaum)

A *cascarillo* strips bark from a stand of cinchona trees on "Spruce's ridge."
(Copyright © Mark Honigsbaum)

Gathering and drying cinchona bark in a Peruvian forest, circa 1867. (Wellcome Library, London)

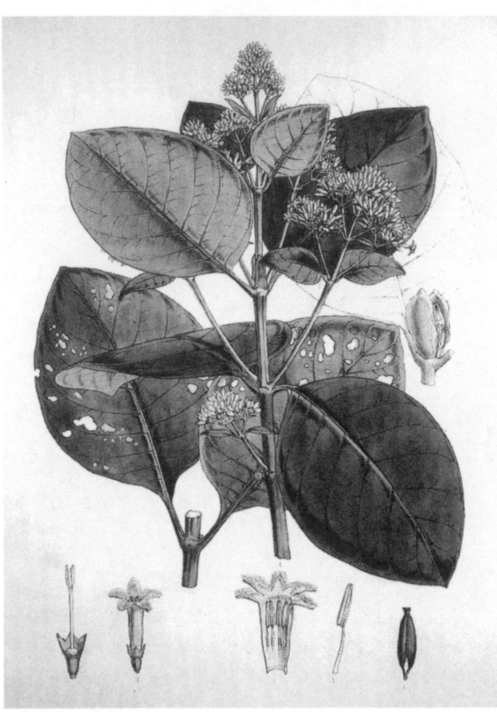

A colored lithograph of *Cinchona succirubra* from John Eliot Howard's *Illustrations of the Nueva Quinologia of Pavon*, 1862. (Wellcome Library, London)

The governor of Madras planting the first cinchona tree in the Nilgiri Hills, India, circa 1861. (Wellcome Library, London)

6

The Quest

The gods, that mortal beauty chase,
Still in a tree did end their race:
Apollo hunted Daphne so,
Only that she might laurel grow:
And Pan did after Syrinx speed
Not as a nymph, but for a reed.

—ANDREW MARVELL,
"Thoughts in a Garden"

BY THE END OF 1859 the Dutch were beginning to despair. After arriving in Java, Hasskarl had planted his *Cinchona calisaya* seeds in a volcanic range to the south of the capital, Batavia. But the soil was thin and the site ill-suited to cinchona, and the plants had become more and more sickly. However, instead of cutting their losses and propagating a more promising variety, the Dutch simply moved the plantation to the Malawar Mountains.

This was a pity. In 1856 Hasskarl's successor, Franz Junghuhn, had arrived in Java with two *C. calisaya* plants raised from Weddell's seed. Unlike Hasskarl, Weddell had personally collected the *calisayas* in Bolivia, but because they were less hardy and more difficult to cultivate, the Dutch put less effort into their propagation. The result was that three years later the Dutch had nearly a million of Hasskarl's *Cinchona pahudiana*—as his variety of *calisaya* became known—growing in the Malawar Mountains, and just seven thousand of Weddell's *calisayas*. It was only when the Dutch began testing the bark for quinine, however, that they realized their mistake. De Vrij, the plantation's chemist, was astonished to find that the *pahudiana* contained just 0.4 percent of the alkaloid, whereas Weddell's *calisayas* contained up to 5 percent. Indeed, so minuscule was the quinine in Hasskarl's plants that it was doubtful that they were true *calisayas* at all. Clemente Henriquez, the Bolivian he had hired to collect the seeds and plants, was emerging as a man of considerable complexity. Not only, as Ledger attested, had he poisoned the soil in Hasskarl's Wardian cases, but it now looked as though he had performed a subtle sleight of hand with the seeds, too.

Markham was determined not to make the same mistake. The British cinchona hunters would not rely on outside agents but would harvest the plants and seeds in person. In addition, Markham arranged for Kew to send two expert gardeners—Robert Cross and John Weir—to nurture the cuttings and watch over the Wardian cases during the return voyage to England. The idea was that Cross would rendezvous with Spruce at Guayaquil once his seeds were ready to be sent to the coast, while Weir would accompany Markham into the *calisaya* forests of Bolivia and Peru. Markham also arranged for a fifth collector, by the name of Pritchett, to travel to Huánuco in Peru and gather specimens of the gray barks.

Markham's organizational skills cannot be faulted. However, his political strategy left a lot to be desired. Convinced of the urgent need to gather the cinchonas before the South American forests were exhausted, he made little effort to conceal his aims. Instead, he broadcast the purpose of his mission widely in the naive belief that the Andean republics shared his philanthropic objectives and would abide by the rules of "fair play." The Old World had contributed to the New such priceless gifts as wheat, barley, rice, sheep, cattle, and horses, Markham argued. It was only right that the New World should now grant the Old a product of life-and-death importance: the cure for malaria. Besides, demand for quinine would always exceed supply. The Indian plantations might lead to a slight fall in the world price of quinine, but the knowledge gleaned from the cultivation of cinchonas by expert European botanists could only benefit South America in the long run.

> Hitherto they have destroyed the cinchona trees in a spirit of reckless short-sightedness, and thus done more injury to their own interests than could possibly have arisen from any commercial competition . . . It will then be a pleasure to supply them with the information which will have been gained by the experience of cultivators in India, and thus to assist them in the establishment of plantations on the slopes of the eastern Andes.[1]

What Markham appeared to have forgotten is that cinchona was an essential part of Peru's national identity. When Peru gained independence from Spain in 1821, a cinchona tree was enshrined in the national coat of arms along with an alpaca and a horn of plenty filled with gold—images that were meant to underline the agricultural, animal, and material wealth inherited from the Incas.[2] Not only that, but the Condesa de Chinchón's cure had taken place in Lima. Regardless of whether the story was true, it was widely believed. Under the circumstances it was naive in the extreme of Markham to think that the Peruvians would relinquish such an essential part of their botanical heritage without a fight. Unfortunately, by the time he realized this, it was too late: as in the days of the Incas, when imperial runners, fueled by coca, conveyed messages the length of the Andes with astonishing speed, so news of Markham's expedition preceded his passage over the cordillera.

MARKHAM SET SAIL on his quest on December 17, 1859, accompanied by his wife, Minna, and John Weir. His first stop was Guayaquil, where he planned to make the arrangements for the collection of Spruce's plants and seeds. But Ecuador was "in a complete state of anarchy," and with the Peruvian fleet moored in the estuary, Markham found himself negotiating directly with the president of

Peru. Fortunately, Walter Cope, the British consul, had already obtained passports guaranteeing the safe passage of Spruce's mules to the bark forests. But the political situation was apt to change at any moment, and Markham left strict instructions for Cross to load the seeds and plants onto the first steamer for Southampton as soon as he arrived in Ecuador.

Lima, in contrast, was an oasis of calm. Disembarking on January 26, 1860, Markham was greeted as a homecoming son. He was now an eminent Inca scholar, and the friends he had made during his travels seven years previously were delighted to see him and his new wife. One writer, reviewing Markham's book *Cusco and Lima*, could barely contain his enthusiasm. "Mr. Markham does not come alone this time," he gushed. "He is accompanied by a young and beautiful wife, who participates in his enthusiasm and love for the privileged land which they now come to explore; and she is herself gifted with brilliant intelligence, and the most amiable qualities."[3]

Markham's reaction to the review was characteristically naive. "This has, I think, thrown people off the scent as to my real object, although it must necessarily be known to several," he confided in his diary. Indeed it was.

Markham spent a month in Lima organizing supplies and equipment and discussing the condition of the cinchona forest with landowners in the eastern cordillera. Their prognosis was not encouraging. Ten years earlier, they informed Markham, it would have been relatively easy to harvest the plants. But in Caravaya, true *calisayas* were now so scarce that collectors had to cut deeper and deeper into the forest. Conversely, on the Bolivian side of the border the forests were still rich in *calisayas*, but war with Peru was imminent and the *cascarilleros* were more suspicious than ever of foreigners.

The intelligence placed Markham in a quandary. Should he proceed to Bolivia and risk running foul of the *cascarilleros* or, worse, being trapped behind the border by the outbreak of war? Or should he play it safe and go to Caravaya where, like Hasskarl, he might find only inferior seed?

The whole success of the mission, not to mention Markham's place in history, depended on his making the right decision, and he was understandably nervous. "The forests where the cinchonas grow are almost inaccessible, without roads or habitations. There is great risk of being disabled by fevers in the humid forests, by *sorochi* [*soroche*, altitude sickness] in crossing the Andes, and by dysentery and ague on the coast. The plants, after they are procured, will have to pass on mules' backs over the frozen planes of the Andes, and the hot arid deserts on the coast before reaching the Wardian cases . . . Many of these obstacles may, of course, be overcome by energy and care, others are in the hands of Providence."

AT THIS STAGE Markham was still hoping to reach La Paz. Today this is a relatively simple matter. From Islay on the coast of Peru there is a railway all the way to Puno on the shores of Lake Titicaca, from where you can catch a hydrofoil and bus to Bolivia's capital. In 1860, however, there was no railway and there were no roads—just a mule track that led across Peru's coastal desert to Arequipa and then wound up the cordillera and over the paramo—the high, treeless plateau—to Puno. Crossing Lake Titicaca, the highest lake in the world, was equally problematic. In Markham's time the only vessels that plied the freezing cold waters were balsas, light reed sailboats. Although they were supposedly unsinkable, progress was necessarily slow, and the balsas were limited in the loads they could carry. And once the eastern side of the lake was reached, the problems were by no means over. To reach the bark forests below Apolobamba, Markham faced an arduous expedition across a series of snow-swept plains and down precipitous mud tracks.

Nevertheless, in early March he loaded his mules and set off across the coastal desert accompanied by Minna and John Weir. But en route he suffered a severe bilious attack, and by the time they reached Arequipa he was very weak. At this point he decided to part company with Minna, fearing that if he did cross into Bolivia, her life might be in danger. "I cannot fully decide until I reach Puno," he wrote in his journal, "but I am fully resolved never to leave the coast of Peru, without a supply of *Calisaya* plants. With God's blessing I *will* succeed!"

Although it pained him to leave his "beloved wife," Minna pointed out that from Arequipa she could pass on his communications to the other cinchona hunters and keep him informed of political developments. Thus, on March 23, Markham started for Puno along the same road he had taken in 1853.

His route took him toward Cusco before veering east up the *alto de los huesos*, a long, zigzag path to the cordillera. It was the worst season of the year to be traveling, and they were soon being pelted with hail and icy rain. On reaching Apo, at fourteen thousand feet, the majority of the party, including the mule drivers, began complaining of *soroche*—"violent pressure on the head, accompanied by acute pain and aches in the back of the neck." In the night, their symptoms intensified, and by morning Markham hardly had the strength to mount his mule.

He was now in the most desolate country he had visited since his voyage to the Arctic in search of Sir John Franklin ten years previously. Only the herds of vicuña roaming the mountainside and the sight of a *coraquenque*, an Andean condor, reminded Markham he was in South America. Then, no sooner had they reached the Alto de Toledo, at 15,590 feet the highest part of their route, than they were forced to cross a series of bogs. With each tentative step the swampy mud threatened to swallow their mules. Eventually, however, they reached the swollen banks of the Tortorani River and were greeted with a dazzling view of Lake Titicaca and the snow-capped peaks of the Bolivian Andes behind it. By the time Markham reached Puno on March 27, he was exhausted. Almost immediately he discovered that the Peruvians had stationed three divisions near the Bolivian border and were anticipating war. Not only that, but the rains had been excessive, turning the roads on the eastern side of Lake Titicaca into mud chutes. Given the Bolivian ban on the export of *calisaya*, Markham concluded that it would be "next to impossible for a private person to preserve his mules from seizure" and that it was altogether too risky to cross the border. Caravaya, on the other hand, lay in Peru and was nearby. Contrary to his earlier thinking, Markham now convinced himself that if he penetrated far enough into the Peruvian forests, he would still be able to find young *calisaya* trees in fruit. Moreover, because very little bark was harvested there, he reasoned that his party would not arouse the jealousy or suspicion of the natives. In fact, this proved another example of his wishful thinking.

WHILE MARKHAM WAS PONDERING his next move, Spruce had been doing his best to ignore the civil war raging in Ecuador. Through discussions with local traders he had discovered that the best red bark did not come from Alausí at all, but from the cinchona forests in the western foothills of Chimborazo. Covering an area of two thousand square miles, the forests were the property of two men. One, Dr. Francisco Neyra, was a local notary, but by a stroke of luck the second was General Flores. Flores had recently been recalled to Ecuador by the provisional government after an absence of fifteen years to help take back Guayaquil from the rebels. Spruce now contacted James Taylor, the general's former doctor, to talk about renting his land. At first, Flores was reluctant, but Spruce persevered. The trees were about to blossom and, he insisted, he had to be on the spot to gather the seeds when they began to form. Hurriedly, he sent word to the British consulate in Guayaquil that he needed money to clinch the deal. Cope duly forwarded the funds entrusted to him by the India Office, and after lengthy negotiations, Flores and Neyra agreed to let the forest to Spruce for $400. The only condition was that he leave the bark intact.

Neyra now became very cooperative, offering Spruce the use of his *cascarilleros* from Guanujo, a town just beneath Chimborazo. Through Neyra, Spruce also learned that the most highly valued bark of all came from Limón, a ridge at four thousand feet surrounded by cane farms. Although the largest trees had been cut down some years before, the *cascarilleros* had carefully nurtured the new shoots, and Spruce was informed that the young trees were now stout enough to bear flowers and fruit. Spruce also needed to collect specimens of *Cinchona officinalis* from Loja, but Loja was separated from Limón by a fifteen-day journey. Realizing he could not be in both places at once, he rode to Riobamba to persuade Taylor to collect the *C. officinalis* on his behalf. Taylor agreed, and Spruce returned to Ambato to rest and await the arrival of the dry season.

Ecuador was now divided into two opposing armies, with the rebels controlling Guayaquil and the low country, and the provisional government, Quito and the sierra. But the threat of being swept up into one or the other of these warring factions was nothing compared with the challenge of crossing Chimborazo. The dry season coincided with the Andean winter, when conditions on the "glory of the Andes" were well below freezing. To reach Limón, Spruce would first have to ride to Mocha and ascend the northwest shoulder of the volcano to as high as thirteen thousand feet. Then he would have to work his way down a precipitous, muddy slope to six thousand feet along tracks so narrow that even mules found it difficult to keep their footing. Of course, Humboldt and Bonpland had managed it sixty years previously, climbing to 19,286 feet—the highest point at that date reached by man. But even for someone in peak physical condition the mountain was a formidable challenge. How would Spruce, always suffering from one ailment or another, cope with such an arduous ascent? The question must have been playing on Spruce's mind, or perhaps it was simply a case of bad luck, but he now suffered a mysterious and debilitating attack. His journal entry of April 29 is certainly melodramatic. Headed "The Break Down," it reads: "Woke up this morning paralysed in my back legs." Some time later, it appears, he went over the entry in black ink, underlining the words *Break Down* twice and adding in parentheses, "From that day forth I was never more able to sit straight up, or to walk about, without great pain and discomfort, soon passing to mental exhaustion."

Spruce later described his condition as "a rheumatic and nervous affection, almost amounting to paralysis." While he could still move about and make preparations for the journey, he now considered it doubtful that he "should be able to bear the arduous fatigues of travelling in the forest" and decided to delegate his commission to Taylor. Was this an extreme "case of the vapors," brought on by the pressure of the impending mission, or was Spruce genuinely ill? One suggestion is that he may have been suffering from Guillain-Barré syndrome, a rare allergic reaction of the peripheral nerves usually brought on by a severe respiratory or gastrointestinal infection. Certainly his condition mystified doctors for years. According to Wallace, in 1869 doctors diagnosed the cause of the paralysis as a "stricture of the rectum"—what today we might call a bad case of piles. But that hardly seems sufficient to explain symptoms as extreme as "paralysis."[4]

ON THE FACE OF IT, Markham was made of sterner stuff. While Spruce was convalescing in Ambato, he had reached Crucero, the capital of Caravaya, and begun descending the eastern cordillera to Sandia, the town where Hasskarl had arrived disguised as a German six years previously. Two days from Crucero, he arrived at a hut at the foot of a long cliff and met a ruddy-faced man who gave his name as Don Manuel Martel. A former colonel in the Peruvian army, Martel had lost a large amount of money in the bark trade. Now aged fifty-two, he claimed to have met Hasskarl in 1854 when Hasskarl was traveling under the assumed name of Muller. He said that Hasskarl had offered him $40 to collect cinchona seed, to which he had replied, "Not for $500." Martel then pointedly told Markham how Hasskarl had hired Henriquez and how Martel had vowed that if Hasskarl or anyone else ever again attempted to take cinchona plants out of Peru, Martel would stir up the people and cut their feet off. "There was evidently some allusion to myself in these remarks," writes Markham, "and I suspected, what afterwards proved to be the case, that Martel had, by some means, got information respecting the objects of my journey, and was desirous of thwarting them."

Markham's suspicions would be proved right at Sandia. Forgoing a muleteer, he had had a frustrating journey to Crucero and across the thirteen-thousand-foot pass that lay above the town. His pack mules had bolted at every opportunity, but rather than invest in his own mule train he had opted to exchange the animals at post houses en route. This kept his costs low but also guaranteed that he was encumbered with the meanest and most recalcitrant beasts. The journey down from the pass had been particularly nerve-racking, as he had had to descend seven thousand feet in just thirty miles. However, it was worth it for the sight that greeted him on arrival. Surrounded on all sides by mountains, Sandia is nestled in a steep ravine over which pour a series of spectacular waterfalls. The stepped hillsides produce some of the finest coffee and coca in all Peru. Moreover, it was in the valley of Sandia that Weddell first encountered the diminutive cinchona shrub he named *Cinchona josephiana* in honor of de Jussieu.

Markham's original plan was to make a leisurely survey of the cinchona forests, locate the most promising stands, and then bide his time until August, when the trees would be ripe and he could gather the seeds. After only two days in Sandia, however, he discovered that Martel had written to local elected officials, warning them of his intentions and urging them to stop him. As if that were not bad enough, Martel had also begun stirring up other villages deeper into the cinchona forest. Markham's mission was fast becoming "the talk of the whole country." Unless he moved quickly, he might not reach the coast alive.

Markham calculated that once he left Sandia there would be no possibility of finding fresh food and provisions, so he loaded up supplies that he hoped would last a month. In those days, explorers did not have the luxury of high-calorie powdered dinners or sophisticated water-purification systems. The food was simple—bread, cheese, chocolate, and concentrated beef tea—and you drank river water. In addition, Markham brought candles, ponchos, powder and shot, a tent, an air bed, and a change of clothes. There would be seven in his party—himself, Weir, a young mestizo boy, Pablo Sevallos, and four Indians to carry the leather bags packed with supplies. No sooner had he left Sandia on the afternoon of April 24, however, than one of the Indians deserted, leaving just three men to carry the provisions.

The trail led down a ravine along a series of ledges overhanging a fast-flowing river. It was narrow and dangerous, but the scenery was magnificent and the vegetation became richer and more tropical with the descent. Markham was encouraged by the fact that the few scattered huts they passed possessed no doors—testimony to the honesty of the people in the region. To keep up with his Indian porters, Markham had adopted the native habit of chewing coca. Like many an explorer since, he found it produced an "agreeable soothing feeling" and meant he could abstain from eating food for long periods. More important, it enabled him to climb steep mountainsides "with a feeling of lightness and elasticity"—a quality that he thought ought to recommend it to "members of the Alpine Club."

No doubt the coca also added to Markham's appreciation of the dramatic scenery. Tier upon tier of coca terraces rose up one side of the ravine, the coca's delicate green leaves contrasting with the deeper hues of the coffee plants, while on the other side the walls plunged dramatically down to the river. It was here that Markham encountered the first wild cinchonas since his journey to the Madre de Dios with Father de Revello eight years before—*Cinchona josephiana* with "exquisite roseate flowers and rich green leaves with crimson veins." Unfortunately, there were too few to be worth collecting. Nevertheless, Markham noted down the best-looking specimens so that he would have no difficulty finding them, if necessary, on the return journey.

Next he decided to leave the ravine he had been following since Sandia and descend to the Huari-huari. After camping by the river for the night, he then climbed up a steep ridge and crossed to the next valley. Behind him lay a series of spectacular ridges framed by distant snowy peaks. At last he was within touching distance of his goal—the cinchona forests hugging the Tambopata River. The last European to visit this area had been Weddell in 1846. Markham's success rested on his equaling the achievement of this great botanical explorer. It was a pivotal moment and one that seemed to bring to the fore all his fears and most deeply seated anxieties.

> I am about to penetrate where no European has been before, and no human being for thirteen years, where there is no road, a dense matted forest, to suffer excessive fatigue, hunger, exposure, torments from insects, dangers from the nature of the country, from snakes and other animals, from savage Indians, and from the risk of accidents without the possibility of help: with the certainty of being vilified by those who sit at home and talk, if unsuccessful, and of having my services depreciated if successful. The feeling of having performed my duty manfully and well supports me through all this, and I know that my dear love will look on me with loving pride when I return, and that all true friends will do me justice.

One cannot read this soliloquy without feeling some sympathy for Markham. He knew he was not a botanist and was honest enough to recognize that his previous excursion to the Madre de Dios River hardly qualified him to lead such an important expedition. He was also intelligent enough to realize that from the India Office's point of view he was highly expendable. After all, what was one tax clerk more or less to the great British Empire?

But if his upper lip lacked the requisite Victorian stiffness, he certainly did not show it in public. Instead he put on a brave face for his colleagues. To lessen everyone's anxieties, he decided to turn the expedition into a game, naming his mule Joe, after the *josephiana*, and issuing the members of his party with pretend military ranks. Thus, Savallos was promoted from aide-de-camp to lieutenant, while Markham was to be referred to as commander in chief. The exercise must have triggered the military historian in him for he soon found himself musing on another British seaman-turned-adventurer, Sir Walter Raleigh. "Our characters were, I think, cast in a similar mould. The same aspirations, the same love of enterprise, some of the same faults. Yet it is strange that Raleigh should have spent so much on the search for El Dorado . . . while I am now in perhaps the richest gold region in the world, and scarcely giving it a thought."

MARKHAM'S FIRST PORT OF CALL was a hut on the Tambopata that Weddell had stopped at in 1846. Used as a rendezvous point by the local *cascarilleros*, it was owned by a local farmer, Don Juan de la Cruz Gironda, and ideally situated for forays into the forest. Markham's supplies were already much reduced, but fortunately Gironda agreed to supply him with enough food to get him back to Sandia. He also introduced Markham to an experienced local bark cutter, Mariano Martínez, who had guided Weddell. On May 1 they set off into the dense, tangled forest. Once again the party had swelled to seven. Martínez led the way, cutting a path through the jungle with his machete; the others followed in single file. The ground was choked with creepers, fallen bamboos, and long tendrils that twisted round their ankles and tripped them at almost every step. In many places they had to scramble along the verge of giddy precipices as the Tambopata surged beneath them. By the end of the first day, Markham was exhausted. "In some places we were half an hour forcing and cutting our way through a space of ten yards; and all the time, besides the torments of insects, I had a head ache and pain in the stomach. Looking back it seems like days, instead of hours that we have been forcing our way through the forest, and we have really only gone fifteen miles."

For three days they continued like this, hacking their way through the jungle and scrambling up the riverbank whenever Martínez spotted a stand of cinchonas. The trees generally grew at 450 to 650 feet above the riverbank, and gathering them was tiring. In some places, they had to cling to sheer precipices with only the half-rotten branches of trees as footholds, while in others they had to hack through foliage thick with thorns and hornets. Most evenings, Markham wrote, he would collapse by the campfire, "sodden and wet to the skin, covered with moss and fungus, bitten all over by mosquitoes, my hands sliced in pieces by sharp grass." Adding to their frustration was the scarcity of good trees. Relatively worthless cinchonas like *Cinchona ovata* were everywhere; the *calisayas* were generally found only on the most inaccessible ridges. Not surprisingly, after four days in the forest the Indian porters threatened to mutiny, saying they had been away from home long enough; besides, they had run out of maize and coca. Markham hurriedly sent one of the Indians back to Gironda to replenish their supplies and begged the others—in Quechua—to stay. Luckily, the appeal worked. "If they had deserted," he admitted in his journal, "all our plans would have been entirely frustrated."

The Indian who had been sent to Gironda soon returned with fresh food, but the supplies could not sustain them indefinitely. Markham held out until May 7, but when he saw that the provisions were once again running dangerously low, he had to face reality: the time had come to turn around. They had managed to collect two hundred cinchona plants. After laying the samples between layers of moss and Russian matting to keep them moist, they set off for Gironda's clearing.

On the way back, it began to rain heavily, making their earlier worries about the plants drying out seem ridiculous. Soon they were so wet that their hands were wrinkled "like washerwomen's" and the damp had penetrated their gunpowder. On reaching a sheer precipice, Weir managed to scramble up the side and collect twenty-one more *calisayas*. The following day they were even more fortunate, finding a further hundred and nine of a variety known as *calisaya morada*, and by the time they reached Gironda's clearing, they had more than five hundred plants, more than enough to fill the Wardian cases that were awaiting their return to Islay. All they had to do was pack them in moss and make their way back to Sandia and over the Andes. But on May 11, just as they were about to leave, Gironda received an ominous letter from the *alcalde municipale* of Quiaca, ordering him to seize the plants and arrest Markham and Martínez immediately. The letter had been hand delivered by the son of Don Manuel Martel, the red-faced colonel Markham had met on the road to Sandia. Markham cursed his indiscretion. During their conversation he had made a passing reference to Martínez. Clearly Martel had guessed that, like Weddell, he intended to hire Martínez to guide him to the cinchonas and was using the fact to create difficulties for Gironda in the hope of getting his hands on his property.

Not surprisingly, the letter filled Gironda with alarm and he immediately proposed disposing of the plants, leaving Markham with just a few, which he could smuggle out without Gironda's involvement. But Markham had come too far and been through too much to give up so easily. Arguing that the *alcalde* had no authority to order his arrest and that he had no intention of harming Peru's national interest, he told Gironda he was leaving immediately and, if necessary, would defend the plants with force. The threat worked, and, after sitting up with the cinchonas all night, Markham set off at dawn the following morning. He was by no means out of danger.

No sooner had he climbed out of the forest than he saw Martel's son and a party of mestizos waiting for him on the *pajonal*—the grasslands. As they drew nearer, Martel's son began cursing the Indians in Quechua, demanding to know how they dared steal Peru's national property and warning them that they would be arrested as soon as they reached Sandia. The situation was poised on a knife-edge. Markham drew his revolver. Martel hesitated. The powder was damp, but Martel couldn't have known that. Slowly, he measured Markham with his eyes; then he stepped back and allowed him to pass. Markham's nerve had held. Now he just had to get to the coast.

Unknown to Markham, however, Don Manuel Martel had been busy rallying the newly formed municipal juntas in the towns that lay along Markham's escape route, and by the time he reached Sandia three days later, an alliance had been formed to prevent the cinchonas' leaving Peru. The Sandians refused to sell him fresh mules and said he would have to go to Crucero, where Martel was waiting for him, to buy them instead. Markham knew full well that that would spell the end of his mission. Fortunately for him, one of Sandia's councillors was sympathetic to his plight, and in return for his gun agreed to supply him with the mules he needed, as well as an Indian guide. In the hope of throwing Martel off the scent, Markham dispatched Weir to Crucero and, avoiding the main road to Arequipa, set a direct compass course over the cordillera to Vilque instead.

It was a hazardous route, but the difficulties were tempered by Markham's knowledge that his test was almost at an end. "Tomorrow, I bid farewell to the montana of Caravaya, that lovely golden land, where I have worked harder and with more anxious care, and done better service than I ever did in my life before. The time has been one of intense anxiety and enjoyment—an earnest desire to do the work well, and a feeling of joy at having so great a work entrusted to me, and a chance of performing a really important service to mankind." On the morning of May 17, he saddled his trusty mule Joe and, driving before him two others laden with the plants, set off accompanied by Angelino Paco, his Indian guide. Markham assumed that Paco knew the country, but as they left the valley of Sandia behind, it became clear that he had never taken this route before. Markham had no choice but to rely on his compass. Then, with night coming on and no moon, Markham pitched his tent behind a range of black cliffs and opened his pack. *Disaster*. His food and matches had been stolen in Sandia. All they had to sustain them was Paco's small store of parched maize.

They spent a hungry night huddled together for warmth, and left early the next morning. Their route now led up the summit of the Cordillera of Caravaya and across a desolate, frosty plain populated only by vicuñas and wild geese. Eventually, after eleven hours in the saddle, they reached an abandoned shepherd's hut and spent another freezing night, with the thermometer dipping as low as 20 degrees Fahrenheit. They awoke to discover that the mules had bolted and spent three hours chasing them down. Despite their fatigue, they then rode for another day, going to bed without supper. All the while Markham had to keep a close eye on the mules to prevent them from rolling and crushing the plants. At night, to protect the cinchonas from the frost, he curled up with them under his poncho. Never mind that he was starving and bitterly cold himself. All that mattered was that the plants were safe.

Finally, on the fourth day out from Sandia, they reached a small estancia where, in exchange for the remainder of his coca, an Indian family provided Markham with milk and cheese. The next day he reached a huge unmarked lake populated by immense flocks of flamingos, as well as ibises, ducks, and cranes. As the flamingos whirled up in a long spiral column, their crimson wings and rose-colored necks silhouetted against the azure Andean sky, Markham allowed himself a brief moment of reflection. "It was," he wrote, "one of the most beautiful sights" he had ever seen.

On May 22 they reached Vilque, but Markham could not risk stopping. After gathering fresh supplies, he left promptly the following afternoon. Ignoring a series of gales and icy squalls, he marched rapidly across the cordillera, reaching Arequipa on May 27. Here he was reunited with Minna and, two days later, Weir, who, as he had anticipated, had found Martel lying in wait for him at Crucero. The subprefect had detained Weir but, finding no plants on him, was forced to let him go. With no one to rally local opinion against him, Markham left Arequipa almost immediately, and the plants were safely deposited in the Wardian cases at Islay on June 1. Of the 529 plants he had brought out of Caravaya, just 73 had perished on the cordillera. His problems were by no means over, however. At Islay, the superintendent of the Customs House refused to allow the plants to leave the port without an express order from the minister of finance. This was potentially an even bigger threat than Martel. The president of Peru had already issued a decree prohibiting cinchona from being exported, and Markham calculated that it was only a matter of time before it was forwarded to the authorities in the interior and customs officers on the coast. However, Markham guessed that because of the ramshackle state of the country, a copy of the decree would not yet exist in the office of the minister of finance, and that if he rode to Lima, he might be able to browbeat officials into giving him the necessary papers. "Besides," he wrote, "all clerks in public offices are changed in every revolution, and there are no officials with more than two or three years' experience. My only hope was in their ignorance, for I well knew that their jealousy of foreigners would induce them to seize upon any plausible excuse for prohibiting the embarkation of the plants."

Leaving Weir in charge of the Wardian cases, Markham rode directly to Lima, where he demanded an audience with the minister of finance, José Salcedo. It turned out that Salcedo had formerly been a Colonel of the Horse and had been appointed by the president—an "illiterate but shrewd Indian"—to show his contempt for the other branches of public service. As Markham had predicted, he knew nothing about the order. Nevertheless, Markham was able to use a combination of cajolery and threats to persuade Salcedo to give the necessary permission.

The cinchonas meanwhile had begun to throw out buds and young leaves—proof that they had recovered from the journey across the cordillera—and on returning to Islay on June 23, Markham immediately had them hoisted on board a launch ready for transfer to the steamer *Valparaiso*. However, the plants were still not out of danger. In the middle of the night an attempt was made to bribe the boatmen to bore holes in the Wardian cases and pour boiling water on them. Then, no sooner were the plants at sea than one of the panes of glass on the cases was smashed. Markham suspected a group of Peruvians he had overheard grumbling about foreigners "robbing the country," and threatened to throw them overboard if he caught them tampering with the cases again. The threat did the trick, and the plants arrived intact at Lima on June 26, crossing the Isthmus of Panama soon after. As Markham watched the American coast disappear beneath the horizon, he could barely contain his sense of achievement.

> It could not but be satisfactory to look back upon the extraordinary difficulties we had overcome, the hardships and dangers of the forests, the scarcity of the plants, the bewildering puzzle to find them amidst the dense Underwood, the endeavour to stop my journey . . . and then to see the great majority of the plants bud-

> ding and looking healthy in the Wardian cases. So far as our work in South America was concerned, it had been performed with complete success.

Whatever the Cassandras at home might say, Markham could return to Britain with his head held high. The quest was over. His only regret was that he had been unable to persuade the British government to take the cases straight across the Pacific to India on HMS *Vixen*, which had been lying idle at Callao. Unfortunately for Markham, that penny-pinching would later prove costly.

7

Spruce's Ridge

The sounding cataract
Haunted me like a passion; the tall rock,
The mountain, and the deep and gloomy wood,
Their colours and their forms, were then to me
An appetite; a feeling and a love
—WILLIAM WORDSWORTH,
"Lines Composed a Few Miles Above Tintern Abbey"

IT IS JUST THIRTY MILES as the condor flies from the summit of Chimborazo to the cinchona forests in the western foothills of the Quitonian Andes, but the descents are among the most dramatic in all the Americas. From the rim of the glacier at 19,000 feet, you plunge through freezing clouds and a series of treacherous bogs to emerge on a bleak, wind-chafed tundra—the *arenal.* On the *arenal* the ground is flat and the going easier, but it doesn't last. Soon you arrive at the Ensillada, the "saddleback"—a craggy ridge on the western shoulder of the extinct volcano. From here you plunge down again, into a verdant bowl-shaped valley, and up the other side to another ridge. Now you are poised above the cloud forest at ten thousand feet, looking toward the Pacific Ocean. Enjoy the view. From here on the descents are almost vertical. To reach the zone where the cinchonas grow, you must drop straight down and up and over another ridge, and another, and another, slipping and sliding along trails that are little more than mud chutes. If your pack animals survive and your legs do not give out, you will be exhausted and exhilarated. You will also have descended six thousand feet in just fifteen miles.

Even for someone in peak physical condition the trail is grueling, but for Spruce it should have been next to impossible. While Markham was fighting his way out of the Tambopata valley, Spruce was still recovering from the mysterious rheumatic and nervous condition that had afflicted him in April. By early June he had been confined to his bed for two months. Unfortunately, we do not know what treatment he received or what thoughts went through his mind during this period; his journal from Ecuador is missing, but his sudden disability must have been mortifying. The India Office had entrusted him with the most important botanical commission of his career. To fail would have been forgivable, but to not even try to gather the cinchonas was unthinkable. Somehow, Spruce had to muster the strength and determination that had seen him recover from malaria and other threats to his life on the Orinoco and Amazon—and cross Chimborazo.

SPRUCE HAD ALWAYS DONE what had been asked of him before, but on this occasion it took James Taylor's intervention to effect a cure. On hearing of his friend's illness, Taylor had rushed to his bedside in Ambato and prescribed various remedies. We do not know what they were, but they cannot have been very effective, as he was soon resorting to psychology, telling Spruce that if he could only get over the sierra, he might find the drier Pacific air on the western side of the mountain reviving. Incredibly, the pep talk worked, and on June 11, Spruce hauled himself from his sickbed and set off with Taylor on horseback for Mocha, a small village to the east of Chimborazo, with a view to traversing the southern shoulder of the volcano and descending to Limón—the ridge where the *cascarilla roja* grew. If

Spruce's report of the expedition is to be believed, it was the most demanding he had ever undertaken. "Not to interrupt the rest of my narrative with the continual groanings of an invalid," he writes, "I may say here, once for all, that . . . I was but too often in that state of prostration when to lie down quietly and die would have seemed a relief."

His illness had left him so debilitated that on the first day out he needed two long rests, and by nightfall he and Taylor were still picking their way over the *paramo* of Sanancajas, a treeless plateau on Chimborazo's southern flank. Thick cloud had obscured their view of the mountain during the ride from Ambato, but at sunset the cloud lifted, leaving, Spruce writes, "a slender meniscus, assuming exactly the form of the cope of the mountain," clinging to the summit. At length this too dissipated, exposing the monarch of the Andes in all its "glory," and allowing them to pick their way by the moonlight reflected off the glacier. Although the sight revived his spirits, by the time they reached Chuquipogyo, a small run-down settlement at thirteen thousand feet, Spruce was so exhausted he found sleep impossible and the following morning was unable to mount his horse. Spruce does not record whether he had a headache and nausea—the most common signs of altitude sickness—but he was now at an altitude where stronger men had succumbed, and he took the whole of the next day and night to recuperate. The following morning, however, was fine, with only a few drops of rain and sleet, and Spruce felt sufficiently strong to resume the journey.

Skirting around the edge of the glacier, he and Taylor soon found themselves on the *arenal*. Here the road became flat and smooth, like a "gravel-walk in England." They had now crossed the watershed between the Oriente and Pacific weather systems, and while on the eastern side of Chimborazo it had been rainy and overcast, on the western side of the mountain the sky was clear and a fierce wind buffeted their horses. They were still above twelve thousand feet and the only vegetation was small hassocks of *Stipa* and *Festuca*, but as they descended to the Ensillada they began to see bright yellow calceolarias, red gentians, violas, and other pretty Andean flowers. Then, at two thousand feet below the *arenal*, Spruce came upon the first trees—the birch-colored *Polylepis*, with its peculiar flaking bark, *Podarcapus*, and *Araliacea*, known as tiger's paw because of its hairy, palm-like leaves.

Ecuador was still embroiled in civil war, and at the Ensillada they saw that several low straw huts had been erected to provide shelter for government soldiers preparing to march on the Peruvian-backed rebels at Guayaquil. As they approached, four raw-looking youths armed with lances rushed out and demanded their passports. Spruce considered unholstering his revolver and firing a shot in the air, but then he remembered he had a far better weapon—a bottle of *aguardiente*. The fiery brandy melted the soldiers' opposition, and Spruce continued down, crossing the saddleback and dropping into the next valley. By nightfall, they had reached Guaranda, a lively market town perched, like Rome, on seven hillsides. Here Spruce was detained for several days, buying fresh provisions and pack animals for the forthcoming descent to Limón. So far, Spruce and Taylor had relied on mules, but the trails on the western slopes of the Quitonian Andes are so steep, narrow, and slippery that their provisions now had to be carried in single file by *cabrestillos*—tough Andean bulls. Before leaving Ambato, Spruce had made a deal with Francisco Neyra, the owner of the bark forests, that he would hire eleven of his *cascarilleros* to guide him to the best trees and help him collect the plants and seeds. But at Guaranda he learned that the *cascarilleros* had been pressed into military service and were forbidden to leave.

Neyra secured the services of four Indians instead, but Spruce soon discovered they knew little about cinchonas or their cultivation. Fortunately, by now Robert Cross—the Kew gardener whose job it was to nurse the cinchonas and convey them to India—had arrived at Guayaquil, and Spruce sent him a message, via the British consul, instructing him to rendezvous with him at Ventanas, a village just below Limón.

On June 17, Spruce and Taylor were finally ready to leave. They loaded their provisions of potatoes, peas, and barley meal, making sure the sacks did not protrude too far from the bulls' backs, and set off for Guanujo, where a trail led up the opposite side of the valley to the next ridge and the pass of Llullundengo, a small V-shaped cutting that guards the entrance to the Pacific cloud forests. From here there was only one direction to go—straight down.

According to Spruce, the descent from Llullundengo at ten thousand feet was "the most precipitous and dangerous" he had ever negotiated. The trail was not so much a track as a near vertical mud ladder rutted with camel humps from the constant passage of animals' hooves. In places, the humps were so slick with grease that it was all he could do to guide his horse from one to another without its losing its footing. One false step and Spruce risked tumbling headlong into the clouds.

His goal was a ridge some six thousand feet below the pass, overlooking the river Limón. Even today the descent is both arduous and spectacular. For the first fifteen hundred feet you feel like Jack scrambling down the beanstalk. Then the clouds disappear and the hillside is a riot of color: yellow *Calceolariae*, blood-red *Melastomaceae*, the violet of spreading *Pleroma*, and every now and then a bromeliad camped in the branches of a tree, its fiery tentacles reaching for the sky.

It took Spruce all day to reach his first campsite, some four thousand feet below the pass, beside the Río de las Tablas. But that valley was only the first in a series of troughs. From there he had to climb up and over another ridge and another; the trail became narrower and more treacherous with every descent. At one place, Spruce writes, the track was so deep and straight he had to throw his legs onto his horse's back to save them from being crushed, and at another, the path was so pitted with stones that there was "a risk of both horse and rider turning a summerset."

Gradually, as Spruce neared the cinchona region, the air became warmer and more humid. As he entered the rain forest, ferns and giant palms formed a canopy over his head. Then, at just below four thousand feet, Spruce writes, "we came out on the first *chacras* [small farm] at Limón, where I almost immediately noted, and with no small satisfaction, a group of three red bark trees, each consisting of from two to four stems of 30 feet high, springing from old stools, and bearing a small quantity of fruit."

From here, he had just two miles of gentle descent before arriving at the cane mill that was to be his home and collecting station for the next three months. "*Cinchona succirubra* is a very handsome tree," writes Spruce, "and in looking out over the forest I could never see any other tree at all comparable to it for beauty. Across the narrow glen below our hut . . . there was a large old stool from which sprang several shoots, only one of which rose to a tree . . . This tree was 50 feet high . . . It had never flowered, but was so densely leafy that not a branch could be seen; and the large, broadly oval, deep green and shining leaves, mixed with decaying ones of a blood-red colour, gave the tree a most striking appearance."

There are no trees the same size on Spruce's ridge today—they were all stripped of their bark and felled years ago—but the ruins of the cane mill are still there, and when the mist lifts the view is spectacular. The ridge runs east to west, falling gradually on the northern side to the Limón River, and very steeply on the southern side to a tributary of the Chausan. The top of the ridge has been given over to a banana plantation, but the slopes are still littered with young *Cinchona succirubra*—descendants, perhaps, of the trees described by Spruce a hundred and forty years ago.

Spruce and Taylor found their accommodation rudimentary. The cane mill was merely a long, low shed with rough, ill-fitting planks that let in the chilly mist, and Spruce soon found that his clothes had turned moldy and his cough had returned. From questioning one of the first settlers at Limón, Spruce learned that the forests above the ridge used to be full of bark trees, as well as bears and pumas. At first the settlers had used the trees for firewood, but soon dealers came from Guayaquil offering them money for the bark, and, as they could still keep the firewood, they began felling trees throughout the forest. In no time every ridge and valley had been stripped bare, so that by the time the landowners excluded the bark dealers with a view to preserving the trees, the damage had been done.

Incredibly, the settlers had no idea of the bark's true worth. According to Spruce, the prevalent opinion was that the Europeans used it to make a chocolate-colored dye. When he tried to explain that the bark was the source of the "precious quinine which was of such vast use in medicine," Spruce reports that they scoffed in disbelief, saying to one another, "It's all very fine for him to stuff us with such a tale; of course, he won't tell us how the dye is made, or we should use it ourselves for our ponchos and our bayetas."

Like Humboldt, Spruce observed that the *cascarilleros* also refused to take the bark for malaria, believing that its heat would only exacerbate the fever and that the correct treatment was *frescos*—cooling drinks. The same prejudice even extended to quinine, as Spruce discovered when he tried to offer some to a young mill hand who had fallen ill on the way back from the low country around Ventanas. "I offered to give them quinine, but they had evidently a great horror of it, and had I given it to the lad in the remissions of the fever, they would have mixed nostrums of their own, so that, had the case gone wrong, the blame would have been laid on us."[1]

Spruce could not let such prejudices influence his own work, and no sooner had he set up home in the mill than he began scouring the ridge for viable seeds to ship to India. It did not take him long to find thirty cinchonas in full flower, but when he collected specimens and dried them, he discovered the seeds were not yet ripe. The reason was the damp climate. For most of the day the ridge was draped in a cool, impenetrable mist. It was only in the late morning that the sun broke through the clouds, ripening the seeds, but by early afternoon the mist had rolled in again and the sunlight was gone. At Guaranda the *cascarilleros* had assured Spruce that he would find plenty of seedlings, but he now discovered that they were simply shoots from old stools and that the only young plants that were viable were the suckers of *Cinchona magnifolia*—worthless because the species contains hardly any quinine.

Spruce could scout the virgin forest for older *Cinchona succirubra* trees in the hope of finding young seedlings growing nearby, but most of the larger trees had already been stripped to their root balls, and, without the aid of the *cascarilleros*, the chances of finding untouched stands were low. To succeed in his mission, Spruce needed the help of an expert, so he sent Taylor to Ventanas to look for Cross. But ten days later Spruce received word from the Spanish minister that Cross had been taken ill at Guayaquil and had yet to set out.

Spruce decided to press on with the work. He recalled Taylor, and together they began scouting the ridge, noting the locations of the best trees. The bark forest stretched for about four miles, and since there was a fair amount of sun at the end of June, Spruce had high hopes of the seeds ripening rapidly. But July was cool and foggy again, with the result that most of the capsules scarcely increased in size while others were attacked by maggots and mold. The climate filled Spruce with despair. "I began to fear we should get no ripe seeds this year, and as the seeds had been especially recommended to me in my instructions from England, it may be imagined how severe was my feeling of disappointment."

Now, to add to his worries, the locals decided that he should buy seeds directly from them, and one morning, while he was out making his rounds, he discovered that two of the trees had been stripped of their flower clusters. Concerned that the remaining trees would be ruined in the same way, Spruce toured the local *chacras*, offering a gratuity to the owners to prevent anyone but him from approaching the trees and telling them the seeds would be useless unless he collected them himself. "This had the desired effect," writes Spruce, "and I do not think a single capsule was molested afterwards."

However, his problems were far from over. About the same time Taylor left Ventanas, the government troops garrisoned on Chimborazo began marching down from the sierra to attack the revolutionary forces that held Guayaquil and the low country. Contrary to all expectation, they selected the route via Limón and Ventanas. "For six weeks we were kept in continual alarm by the passage of troops, and it required all our vigilance to prevent our horses and other goods being stolen . . . Indeed if I had had no companion who was independent of the political feuds of the country, I do not see how I could have got on at all."

It was now impossible for Spruce to send Taylor to Loja to collect specimens of *C. officinalis*, because he needed him to ride to Guaranda each month to replenish their precious stocks of barley and pease-meal that, due to the moist climate, rotted rapidly. Spruce could not send his Indian servants for fear of their being press-ganged into the army, and without Taylor, he writes, "I might literally have perished of hunger."

Meanwhile, Spruce still had no idea what had become of Cross—he might have died or he might have left Guayaquil and been on his way to Limón. Spruce could not afford to wait. He began noting the location of healthy trees that had failed to flower that year with a view to striking his own cuttings. He was just about to start out for the trees with Taylor and two Indian guides when he heard that an Englishman had been spotted at Ventanas with a cargo of peculiar boxes. Spruce immediately sent Taylor to Ventanas to investigate. The Englishman was none other than Cross, and the peculiar boxes were the fifteen

Wardian cases that Markham had deposited at Guayaquil the previous December. Cross had had all sorts of obstacles thrown in his way by the troops who held the river. He had scarcely been able to find enough men to paddle his canoes, and the journey from Guayaquil had taken him nearly two weeks. Nevertheless, despite being pale and thin from his recent illness, he was keen to start work, and after resting for one day he began setting up an experimental nursery. Cross had been forced to leave the Wardian cases at Aguacatal, a village near Ventanas, for safekeeping, so to keep the cuttings shaded and free of fungus he built frames and an adjustable roof out of palm fronds. By the beginning of August he had planted more than a thousand cuttings, and within three weeks a number had begun to take root. At first the weather was foggy, but by mid-August the days were scorchingly hot and the only way to keep the sun from drying out the plants was to ferry water to the nursery in bamboo buckets. Then, just as the cuttings were beginning to thrive, they were attacked by caterpillars. Cross dealt with these, too, and, Spruce writes, it was only thanks to his "unremitting watchfulness" that the plants survived.

Meanwhile, the sun had begun to ripen the capsules on the cinchona trees in the forest, and on August 13, Spruce noticed that they were beginning to burst at the base. He quickly ordered his Indian porters to climb the branches and begin breaking off the panicles one by one while he spread a sheet on the ground so that none of the seeds would be lost. In this way, within two weeks all the seeds had been harvested, and by early September they were dry.

The pains taken in collecting the seeds paid off. Cross sowed eight on August 16, and by the first week of September all were producing healthy plants. By now, his rooted cuttings were also thriving, and he had more than enough to fill the Wardian cases. But Spruce, wanting to be sure he had obtained all the best seeds, now rode south across the valley to San Antonio, where Pavon had originally collected the red bark, to take an inventory of the trees.

At San Antonio the cinchona forests extend to a lower elevation than at Limón, reaching almost to the Guayaquilian plain. Consequently, the trees are more exposed to the sun and wind and the town is unhealthier, with fevers a common complaint. Not wishing to linger, Spruce collected five hundred capsules and hurried back to

Limón. He now had a total of twenty-five hundred capsules. As each capsule contained an average of forty seeds, Spruce calculated that he had at least a hundred thousand—more than enough to satisfy the needs of empire.

Now, at last, he learned that the provisional government troops under the command of General Flores had taken Guayaquil and that, as a result, communications with the interior had been reopened. Spruce decided to ride to Bodegas immediately and from there catch a steamboat directly to Guayaquil, from where he could dispatch a portion of the seeds to Kew while awaiting the arrival of Cross with the plants. He reached Bodegas (now Babahoyo) without difficulty, but once there he discovered that the only available steamer had been commandeered by the government. He had no choice but to wait eight days, finally arriving at Guayaquil on October 7. Seven days later Spruce dispatched two packets of seeds on a Panama-bound steamer: one to Kew and Sir William Hooker; the second to the Botanic Gardens in Jamaica.

SPRUCE HAD SUCCESSFULLY COMPLETED the first part of his mission, but concluding the second part would prove far harder than either he or Cross could have imagined. Spruce had left Limón without knowing when the plants would be sturdy enough to bring down to Ventanas. Since it was all but impossible to get a message to Cross, he simply had to sit it out in Guayaquil and await news from his colleague. This was easier said than done. Revolution was still in the air, and hardly a day went by without a fire or some other commotion in the busy coastal port. In addition, Guayaquil lies on a fault line, and it was now hit by a series of violent earthquakes. Fortunately, the wooden house Spruce had rented for himself and his botanical specimens was well built and survived the shocks. Nevertheless, it was a relief when in November he received word from Cross that the plants would be ready by the end of the month.

On November 24, Spruce returned to Bodegas and proceeded upriver to Aguacatal, the village where Cross had deposited the Wardian cases the previous July. Spruce spent the next two days carefully hammering the cases together. Then he cast about for a raft big enough to accommodate the cases and allow space for cooking and sleeping. He eventually found one a little farther upriver. It belonged to an elderly black man and was constructed from twelve huge trunks, sixty-six feet long and one foot in diameter, the largest that could safely navigate the river. The old man agreed to rent the raft to Spruce for $90, and together they brought it down to Aguacatal, where Spruce loaded the Wardian cases ready to receive the plants from Cross.

Cross did not arrive until December 13. It had taken him longer than he expected to construct bamboo baskets for transporting the plants. He had not dared to entrust such a delicate operation to anyone else and had insisted on carefully wrapping each plant in wet moss. The result was that only a few plants were lost in the descent from Limón—a consequence of the bulls running through the bush—and by the time he transferred them to the Wardian cases, he had 637 ready for transport to India. Rather than run the risk of the glass in the cases breaking on the river, he left the windows off, stretching moistened strips of calico around the cases instead to protect the plants from the elements.

By now the rains had begun and the level of the river was rising rapidly. The only means of steering the raft was with long bamboo oars, two in the prow and one in the stern. On December 24 they set off, five in all: Spruce, Cross, two oarsmen at the front, and the elderly black man at the stern. According to Spruce, the river "ran like a sluice." There were so many twists and turns it was difficult to keep the raft in midstream and out of the way of the branches and twiners that hung over the river. After several hours of steering the raft around these perils, the craft suddenly went "dead on" through a mass of overhanging branches. "The effect," writes Spruce, "was tremendous: the heavy cases were hoisted up and dashed against each other; the roof of our cabin smashed in; and the old pilot was for some moments so completely involved in the branches and the wreck of the roof, that I expected nothing but that he had been carried away."

Luckily, the elderly pilot managed to hang on, but the deck now presented "a lamentable sight." Branches and sticks were strewn all over the raft, and the roof was caved in. After running into the bushes twice more, they finally found an area of the bank that was clear of foliage and hove to to make repairs. Miraculously, there were only a few slight cracks in the cases and none of them had turned over. Spruce spent a fitful night awake, with one eye on the level of the river lest it suddenly fall and the raft be swamped by a landslide from the bank. Then, at daybreak, they continued downriver. This time Cross helped the old man handle the oar at the stern.

At Caracoal the river widened and the banks were less overhung with branches so that the going became smoother. At noon, they passed Bodegas, and by nightfall they had reached the tidal part of the river. It was now simply a matter of riding the ebb and flow until they washed up at Guayaquil. But their progress was so slow that on Christmas Day they were still stuck on the river.

Finally, on December 27, they drifted into the estuary and were able to bring the raft into port. Spruce immediately leaped ashore to fetch a carpenter to assist Cross in repairing the cases. Next, they transferred the cases to a launch and had them brought alongside the dock ready to be loaded onto the next available ship.

Three days later a goods steamer arrived from Lima. The ship was not due to sail again until January 2, and then only to Paita in northern Peru. From there, however, Cross could catch another steamer passing from Lima to Panama. Satisfied that the plants were safely on board and that he had done everything in his power to ensure the success of the mission, Spruce now took his leave of Cross, confident that the plants were "likely to go on prosperously."

LIKE MARKHAM, Spruce believed that in doing his patriotic duty he was also benefiting mankind. But in fulfilling his commission Spruce incurred the wrath of the Ecuadorian government, and in his subsequent report to the India Office he felt obliged to justify his actions.

> I have seen enough of collecting the products of the forest to convince me, that *whatever vegetable substance is needful to man, he must ultimately cultivate the plant producing it* [Spruce's italics]. Whilst the demand for such precious substances as Peruvian bark, sarsaparilla, caoutchouc [rubber] etc must necessarily go on increasing, the supply yielded by the forest will decrease, and ultimately fail. My

> operations in the Red Bark woods have only partially opened the eyes of the Ecuatoreans to this fact, but have in a high degree excited their jealousy at the prospect of another nation participating in the possession of a treasure which they have not known how to preserve from deterioration and proximate destruction, but which, with a little foresight, might have been an endless source of wealth to their country.

Spruce was right about the destruction. By the time he arrived in Limón, the trade in bark was already much reduced. In 1859, the *cascarilleros* had only collected five thousand pounds of red bark, and in 1860 none at all had been gotten out from Limón. The reason for the decline, of course, was the *cascarilleros*' ruthless felling of trees. In the valleys of the Chausan and Limón, Spruce had seen only two hundred mature trees left standing. Of these, only a few had attained their full grandeur of eighty feet, the rest being fifty feet or under. Nearly all the trees were growths from old stools. He was unable to find a single plant that had seeded from a flowering tree in the forest, and in the meadows any seedlings that could have taken root were trodden on by grazing cattle.

On the other hand, in 1859 a quintal—100 pounds, or 45 kilograms—of bark had fetched U.S. $43. Within thirty years, thanks to the success of the Indian plantations, the price had dropped to $10 a quintal—barely enough to pay for the labor of getting it out of the forest. Markham's defense of the missions reiterated how much the Andean republics would benefit from the British example in India, which he hoped would lead them to adopt better cultivation methods. In fact, as we shall see, this never happened. Instead, the success of the Indian and Dutch plantations effectively killed the Andean bark industry. In this respect at least, Spruce's argument that it was impossible to stop men propagating economically valuable plants was more realistic than Markham's.

The Ecuadorians, however, were not nearly as forgiving of Spruce as the Peruvians had been of Markham. Although Spruce had not broken any law, within four months Ecuador had passed a decree forbidding foreigners from exporting plants, cuttings, or seeds and compelling the *cascarilleros* to plant around each felled tree six more plants or shoots of the same variety. The decree also forbade further exportation of the bark, requiring that in future quinine be manufactured on the spot, and inviting Bolivia and Colombia (Peru—understandably, given the recent animosities—was not included) to follow suit.

It was hopelessly unrealistic, of course. Spruce scoffed that the whole thing was framed "on the dog in the manger principle" and that "sticking six branches in the ground" was hardly a scientific way of seeking to restock the bark forests. Nevertheless, in extracting the red bark from Limón, Spruce—like Markham in Peru—had taken advantage of the confused political climate. Had the civil war finished a few months earlier, it is unlikely he would have been allowed to export any samples of cinchona—whether plants or seeds.

ROBERT CROSS had no such excuses. As soon as the *Cinchona succirubra* seeds and plants were safely delivered to India, he returned to South America to collect specimens of *Cinchona officinalis* from Loja. By now it was autumn 1861 and the Ecuadorian decree prohibiting the export of cinchona was in full force. Spruce had originally intended to send Taylor to Loja but had been forced to change his plans at the last moment because of the civil war. Markham, however, was determined to collect specimens of *all* the commercially significant barks, and it was on his instructions that, on September 17, Cross left Guayaquil in an open rowboat and landed at Santa Rosa, the port for the province of Loja. From there he crossed a series of swamps and steep ridges, reaching Loja ten days later. On October 1 he began ascending Cajanuma, the mountain visited by La Condamine and Humboldt. Setting up camp near the summit in a hut belonging to a *cascarillero*, he began exploring the woods for *C. officinalis* stands. After several days of fruitless searching he suddenly spotted a number of fine young trees in a steep ravine. The trees were in flower, and on closer examination he discovered that some of the seeds were ripe. Thereafter, he continued searching all the ravines from sunrise to sunset, often rappelling to the bottom by means of overhanging vines. By the time he returned to Guayaquil in December he had amassed nearly a hundred thousand seeds.

Eighteen months later Markham called on Cross again—dispatching him this time to Colombia to collect specimens of *C. pitayensis* and the so-called Santa Fe and Cartagena barks—respectively, *C. lancifolia* and *C. cordifolia.* Leaving his home on the western slopes of Chimborazo, Cross proceeded directly to Popayán, which lies just across the Ecuadorian border in southern Colombia. From there he headed east to Sylvia, the headquarters of the bark collectors near the volcano Purace. At Pitayo, on the forested slopes and ravines beneath the volcano, he collected 450 ripened capsules of *C. pitayensis,* returning with them to Guayaquil in November.

By now, Markham writes, Cross was "prostrated by ague," but it was not the end of his exertions. In 1868, Markham sent him back to Pitayo to collect specimens he had missed on his first trip. Then, in 1877, he called on him a fifth time, sending him to Caqueta and the Magdalena valley to collect the Santa Fe and Cartagena barks. Cross was once again attacked by "swamp fever" and was only able to keep up with his Indian porters by the continual consumption of coca. But in March 1868, after innumerable sufferings and hardships, he returned to England triumphant.

In 1860, Markham had also succeeded in gathering specimens of the so-called gray barks from Huánuco, in northern Peru. In all, by 1864 he had collected cinchonas from four of the five regions he considered the most important—Pitayo, Limón, Loja, and Huánuco. He had also collected *calisaya* trees from Carabaya in southern Peru, of course, but without the so-called *calisaya vera* of northern Bolivia, the set was incomplete. Unknown to Markham, however, Charles Ledger had been following his activities closely and now began to hatch his own plans.

On arriving in Australia in 1853, Ledger had gone straight to New South Wales, where he presented his proposal for the import of alpacas to the government. According to him, officials gave him a verbal undertaking that if he succeeded in landing a hundred of the animals, his expenses would be reimbursed and he would be granted ten thousand acres on which to graze them. Convinced that the "dry" air of New South Wales was similar to that of the Peruvian highlands, Ledger hurried back to South America to be reunited with his flock, landing at Valparaiso in July 1853. With Peru and Bolivia at war, Ledger realized he could take advantage of the political confusion to move his flock, which was now in southern Bolivia, across the border to Argentina. From there he could drive them along the cordillera to Laguna Blanca, across the Chilean border, and down to Puerto Caldera on the coast, from where they could be shipped to Australia.

It took him nearly five years. Accompanied by Mamani and his son, Santiago, Ledger rejoined his flock at San Cristóbal in November and began driving them to the Argentinean border. Of the 450 alpacas that had set off from Ledger's estate in Chulluncayani, Peru, the previous year, there were now just 312. Conditions were atrocious. It was midwinter, and in crossing the cordillera they found themselves mired in deep snow and ice. As the weather worsened, many of the animals perished in snowstorms, forcing Ledger to stop and purchase fresh alpacas from Indians along the way. To add to his worries, Ledger had run out of money and at one point had to ride a thousand miles back to Valparaiso to raise fresh funds from his backers. Next, he learned that the Bolivians, angered by his crossing of the Peruvian border without a proper license, had issued a warrant for his arrest. He hurried across the border to Argentina, giving the slip to a soldier who had been sent from Potosí to arrest him.

By spring 1856, Ledger had reached Salta, but just as he was contemplating a summer crossing of the Andes to Chile, his flock suffered a massive infestation of leeches. Within days, two hundred animals had died and he was back to square one. Returning to Bolivia to replenish the alpacas was out of the question. Instead, Ledger decided to buy llamas and wild vicuñas and crossbreed them with the alpacas until he had reconstituted the herd. The effort took him the better part of a year. In May 1857, he wintered the herd at Laguna Blanca, a freezing lake in a valley in the high tablelands west of Tucumán. Walled corals were built to contain the animals, and the long days were taken up with nursing the crossbreeds and ice-skating. The surreal scene was well captured by Mamani's son, who, signing himself Santiago Savage, depicted the activities in a series of vivid watercolors.

By May 1858, Ledger had a herd of 619 alpacas, llamas, vicuñas, and crossbreeds, and he was ready to drive the animals to the Chilean coast. It was a journey he would remember for the rest of his life. It took him eight days to cross the Andes into Chile, climbing through a series of passes as high as eighteen thousand feet and sleeping at night on sheets of ice. Then he had to lead the herd down the western cordillera and across the parched Atacama Desert. By the time he reached Copiapo near the Chilean coast a month later, he had lost a further 179 animals and the herd numbered 440. Nevertheless, it was a Herculean feat. The townspeople of Copiapo had never seen alpacas before, and for several days the city was deserted as they filed out to view the herd.

When Ledger finally reached Australia in November 1858, the *Sydney Morning Herald* praised his "unparalleled daring," opining that he had accomplished a feat that had "hitherto been considered altogether impracticable." Unfortunately, his attempts to put the alpacas on a commercial footing were doomed. First, the New South Wales government reneged on its verbal promise, refusing to reimburse his expenses and forcing him to sell the flock to the state legislature to pay off his creditors. Then, no sooner had the legislature "rehired" him to tend to the herd than it began pressuring him to sell the animals off in lots. The problem was competition from the Australian merino sheep, a far more profitable and practical animal than the alpaca and one that was perfectly adapted to the dry Australian climate. Nevertheless, Ledger persevered with his scheme, and in 1862 his alpacas were exhibited at the International Exhibition in London, winning the gold medal. Ledger had to remain in Australia to lobby the bureaucrats, so his brother George collected the award on his behalf. Unfortunately, Ledger's efforts came to nothing, and the following year the flock was auctioned off, ending for good his dreams of an Australian alpaca industry.

Ledger was now broke. He had invested £7,000 of his own money in the alpaca venture and had lost it all. In addition, the New South Wales government was refusing to pay him three months' back salary. There was nothing to do but return to South America. But to what purpose?

Three years previously, while negotiating with Australian officials, Ledger had learned of Markham's expedition to collect cinchonas. Keen to aid the missions, he had written Markham a long letter, advising him how to equip himself, how he should distrust everybody—

even seemingly friendly priests—and where he should go in Bolivia to find *calisayas* of the same quality he and Mamani had seen during their expedition across the Peruvian border in 1851. Ledger had sent the letter to friends in Tacna with instructions for them to hand it to Markham when he next passed through. But by the time Ledger's letter arrived, Markham had already left for Tambopata. He did not return to Tacna but proceeded directly to the Peruvian coast with the cinchonas, so he never received it.[2]

Stuck in Australia, embroiled in the alpaca affair, however, Ledger did not know whether Markham had seen the letter. It was not until 1861 that he had his next news of Markham's mission. Ledger was in Sydney making preparations for Santiago's return to South America when he read in the *Home News* of the arrival in India of the cinchona seeds and plants collected by Markham in Peru. The article revealed that during his search for cinchona Markham had failed to enter Bolivia. Ledger quickly realized that in that case Markham could not have obtained *calisayas* of very good quality—certainly not trees equal to the ones he and Mamani had seen—and thus that his mission was, at best, only a partial success.

Ledger now began hatching a scheme to revive his fortunes. As Santiago was returning to Peru, Ledger would give him a letter to deliver to his father. Recalling all the details of their expedition ten years previously—the route they had taken through Apolobamba, the trail they had cut through the cloud forest with their machetes, and finally, the spot where they had stumbled on the fifty huge *calisaya* trees—Ledger instructed Mamani to return to Bolivia and collect as many seeds as possible. Once this was done he was to return to Tacna, by which time Ledger hoped to have arrived in Peru, and deliver the seeds to him in person.

Ledger's request smacked of desperation.[3] Retracing their steps through the *calisaya* forests was no small undertaking, especially for one man traveling alone. But Ledger was confident that if anyone could do it, it was his faithful Indian servant. The morning Santiago's boat was due to set sail, he handed him the letter together with two hundred Spanish dollars (£40)—the last of his money. Mamani should treat it as a down payment, he told Santiago. When he returned to

Tacna with the seeds, Ledger or his brother-in-law would be waiting with a further £500.

A few hours later, Santiago set sail for Peru with two other Indians who had accompanied Ledger on his epic journey with the alpaca and who had stuck by him throughout his troubles in Australia. For Ledger, it was a last desperate throw of the dice.

8

The Shadow in the Forest

Old Manuel waited patiently, year after year, cutting bark with his sons and looking for an opportunity of fulfilling Mr Ledger's commission.

—CLEMENTS MARKHAM,
Peruvian Bark: A Popular Account

BURIED DEEP IN THE ARCHIVES of the Pharmaceutical Society of Great Britain is a faded photograph of a South American Indian. The man is in his late twenties or early thirties and is standing on a rough path. He is wearing jeans, a sombrero, and the striped poncho favored by the highland Indians of northern Bolivia. A starched shirt and cravat are just visible above the poncho's V, suggesting that he has dressed especially for the occasion. Yet either by accident or design, the photographer has chosen to show him with his hands raised in front of his chest, clasping the strings of his *mochila* (backpack). Indeed, were it not for the Labrador seated faithfully at his feet, you would think from his stern expression that he was about to stride purposefully out of the frame and out of the story forever.

The date on the photograph is 1860, and it could be either Manuel

Incra Mamani or his son, Santiago.[1] We know very little about Mamani, other than that he was from Coroico in the Yungas, and that he was an Indian of either Aymara or Quechua descent. Although Mamani served Ledger faithfully for more than twenty years, Ledger does not give more details or, perhaps, does not care to. This is a shame, for it is thanks to Mamani that Ledger obtained the species of cinchona with the highest quinine content of all. More than Ledger, Markham, or any of Markham's "fellow labourers," it is Mamani who deserves to be remembered for making quinine widely available to the world.

The paucity of information about Mamani's character and motivation has traditionally presented historians of cinchona with a dilemma. Should he be seen simply as Ledger's amanuensis—an Indian faithfully doing the bidding of his "master"—or was he a self-actualizing and conscious actor in the quest? The Spanish historian Marie Louise de Ayala Duran-Reynals controversially took the latter view. In her book *The Fever Bark Tree*, written in 1946, she argues that Mamani's is "a narrative in the best Homeric traditions" and should be seen in the context of the centuries-long subjugation of the Aymara people by first the Incas and then the Spanish. According to Duran-Reynals, it was because "destiny had appointed [the Aymaras] as guardians of the *Cinchona calisaya*" that no white man could hope to wrest the tree from Bolivia without their cooperation. Mamani, however, was "one of the rare Indians of highland Bolivia who had left the native soil and, rarer still, had become the servant of a foreigner." His decision to harvest the *calisaya* for Ledger, she suggests, was a betrayal of his people—but one that he took consciously and willingly out of his "love of the master who had been kind to him."

At the other extreme is Sir Clements Markham. In his *Peruvian Bark*, a popular account of the introduction of cinchona to India, Markham prefers to portray Mamani as Ledger's "true-hearted and faithful Indian servant." In his narrative, Mamani's deeper motivations go unexplored. Instead he assumes that Mamani was simply obeying orders. At one point he even refers to him overly familiarly as "old Manuel" before adding somewhat patronizingly that "it is very pleasant to have to record these noble traits of character in the Indians, the descendants of men who formed and organised the glorious empire of the Incas."

In fact, no one knows whether or not Mamani was a Quechua, and there is certainly no evidence to suggest that he was a direct descendant of Atahualpa, Tupac Amaru, or any other Inca king. More important, there is nothing in Ledger's writings to suggest that Mamani was any nobler than the next man, nor whether his primary motivation was gratitude or money. The truth is, we know so little about Mamani that we can project onto him almost any cultural and historical stereotype we please. He is the shadow in the forest, a cut-out-and-keep acetate for whatever prejudices we care to bring to the story.

There is one theory, however, that neither Markham nor Duran-Reynals seems to have considered, and that is that Mamani was descended from the Kallawaya—the traditional Indian healers of Apolobamba, the highland region of northern Bolivia through which Mamani and Ledger passed while scouting a route for alpacas in 1851. The poncho Mamani or his son is wearing in the photograph, for instance, is identical to the ones worn today by Kallawayas from mountain villages near Bolivia's northern border with Peru. More important, the Kallawaya are expert naturalists. Every winter, when the rains cease, they descend from their mountain retreats to gather medicinal herbs and plants—including cinchona—from the cloud forests below Apolobamba. Then, dressed in their black breeches, striped ponchos, and broad-brimmed sombreros, and with their wallets of herbs strung across their backs, they walk in a direct line from village to village, practicing their healing arts. Indeed, to this day it is said that the varieties of coca and cinchona most prized by the Kallawaya come from Apolo, an isolated jungle town 130 miles northeast of Charazani. There, cinchona is still known by its traditional Aymara name, *calisaya*, from *cali*, meaning "best," and *saya*, meaning "bark"—"best bark." And it was there that Mamani returned, at Ledger's request, for the seeds and plants that would prove the most valuable of all.

REACHING APOLO is no easier today than it was in the mid-nineteenth century. Bolivia's military airline, Transportes Aeros Militares, used to operate a regular weekly flight but canceled it in 2000 because of the poor condition of Apolo's runway and the lack of commercial passengers. Instead, soldiers and civilians wishing to reach the jungle outpost now have to board a bus or truck at La Paz's Cemetario station and endure a grueling forty-eight-hour journey over some of the worst roads in the world. The one that hugs Lake Titicaca, for instance, is little more than an unmetaled track strewn with rocks and pitted with craters. Then, as you plunge from the altiplano, the descents become steadily more precipitous and hair-raising. The only thing that makes the ordeal bearable is the views. First, as you rise up out of La Paz, comes Illimani, its peaks burning like phosphorus in the morning sun. As you begin to cross the altiplano come Huayana Potosí and the Condoriri group, followed in quick succession by Illampu, Ancohuma, and, glinting on the horizon, the Lagunas glacier.

We do not know whether Ledger and Mamani ever traveled this way, but the route must have been well known to them, and in all probability this is the route Ledger recommended to Markham in 1858. Unfortunately, Ledger's letter to Markham is now missing, but Ledger often copied his own writings for different occasions, repeating the same details and many of the same phrases. In a letter addressed to John Howard in 1877, we find Ledger almost certainly reproducing the advice he had given Markham twenty years earlier.

"In the first place," begins the letter, "there are *no laws* prohibiting the collection of plants or seeds, nor against taking them out of the country. There exists, however, the greater prejudice from the highest to the lowest inhabitant against anyone doing it. Were anyone publicly and openly to state his intention, his life would be in the greatest danger." Ledger then goes on to give Howard detailed instructions on how to avoid detection—instructions that would have been invaluable to Markham when he was contemplating a foray into the same forests nineteen years earlier. The collecting party should consist of two men who speak Spanish and who are prepared to "rough it" where necessary. They should pretend to be collectors of bird skins, since when the cinchonas are in flower, the trees are surrounded by brilliantly colored hummingbirds, but they must be careful never to relax their guard, as "every Indian is [a] most observant and suspicious person [and is] very difficult to hoodwink or deceive."

To smooth their passage to the bark forests, Ledger suggests they obtain letters of introduction to local merchants and, most important of all, to the padres at the missions lying along the route. A good start, he says, would be to visit the prior at the Convent of San Francisco in La Paz and "give $25 to $50 to the poor box." Then, proceeding along the eastern shore of Lake Titicaca, they should pass through the towns of Acoraimes and Escoma before veering to the right, from where they can go "direct to Peluchuco, then to Apolo, following on Mission to Mission until arriving on the shore of the Beni."

Ledger concludes: "There is not necessity to encumber oneself with anything . . . Quinine against fevers, ammonia against bites from venomous spiders, snakes (almost all the Indians have special native antidotes) . . . With tact, perseverance and *above all* for no one (not even the Consul) to know one's real object, a man well knowing *when* to pick and how to preserve seed could not fail in his mission."

"A man well knowing *when* to pick": Ledger clearly means to suggest that to succeed a collector would need the sort of expertise that only someone who had spent many years exploring the bark forests could possess—a man, in other words, much like himself. Yet the person who knew most about the trees was not Ledger but Mamani. It was Mamani who could distinguish the different types of cinchona simply by the colors and forms of their leaves, and it was Mamani who had observed the relationship between the topography of the Andes and the distribution of the most valuable quinine-bearing species. Furthermore, as a native of the highlands, Mamani would not arouse suspicion in the same way as a European collector.

Yet even for Mamani such an undertaking was fraught with difficulty and danger. Ten years had passed since he had last seen the stand of fifty huge *roja* trees that had so astonished Ledger. But even if he succeeded in retracing their steps to the headwaters of the Beni River near Apolo, there was no guarantee he would be able to locate the trees again, or that they would be in flower when he got there. Furthermore, if he was lucky enough to arrive just as the seeds were ripening, he would still have to spend several days carefully drying them before transporting them to Ledger in Peru—a process that would inevitably arouse the interest and suspicion of any *cascarillero* who happened to stumble upon him.

Nevertheless, it would seem that no sooner had Santiago delivered Ledger's request to his father than Mamani set off for Apolo determined to fulfill the commission. From what Ledger tells us of their relationship, it would never have occurred to him to do otherwise. After all they had been through together, Ledger and Mamani were no longer master and servant but friends, and Mamani could hardly refuse a friend such an important request.

WHEN LEDGER RETURNED to Peru in 1865, he had no idea whether Mamani had succeeded in locating the fifty huge *roja* trees. It had been a full four years since he had waved farewell to Santiago at Sydney harbor, and in all that time he had not received a single communication from his former guide. Ledger's first priority on reaching Tacna was to visit his three daughters who, because of the alpaca business, he had not seen for twelve years. The reunion must have been difficult, especially because while Ledger had been camped at Laguna Blanca, their mother had died in highly straitened circumstances. Although his father-in-law, Major Ortiz, had taken care of his daughters during his long absence, Ledger now had a new wife—Charlotte Oliver—and was up to his ears in debt. His only hope was Mamani. Had there been any news of his expedition?

To his relief, he discovered that Mamani had already been in touch with Major Ortiz for money three times and had received $200 on each occasion. He had also asked Ledger's father-in-law to give Ledger a message: "Tell the *patrón* that I am only waiting for a good season so as to secure thoroughly ripe seed."

In fact, as Ledger discovered late on the night of May 19, 1865, when Mamani suddenly appeared on his doorstep in Tacna, he had had a very long wait indeed. As soon as he had received Ledger's instructions, he and Santiago had traveled to Caupolicán province, southeast of Apolo, and begun looking for the spot where they had first seen the *roja* trees. They found them, or at least trees very similar to the ones they had seen in 1851, without great difficulty. Unfortunately, it was too late in the season—the Bolivian winter had already arrived and there were no seeds to be had. Mamani had no choice but to retrace his steps and return with his sons the following March, when he would have a better chance of finding the *calisayas* in flower.

This they did for the next *three* years. In 1862 they camped by the trees, cutting bark for their subsistence, and marking the most promising trees when they began to flower. But in early spring there was a severe frost and all the best seeds were destroyed, leaving only inferior sorts to be collected. Incredibly, the same thing happened in 1863, and again in 1864, so that it was not until April 1865 that Mamani had finally been able to collect good seeds.

Mamani described how he had watched the trees become covered in beautiful creamy white flowers like the ones they had seen so many years before. This spectacle was enhanced by the fact that the oldest trees were now covered with a scarlet and silver moss. Then, just before the seeds were about to fall, the undersides of the leaves turned dark purple, completing the brilliant display.

At first he had kept the seeds collected from different trees in separate bags, as Ledger had instructed, but one day while drying the seeds on his poncho there was a sudden gust of wind and they became mixed up. Nevertheless, the seeds were all there and accounted for in the two bags he now presented to Ledger. He had walked all the way from Coroico on foot—some one thousand miles—over the freezing cordillera, sustained only by a little food and coca, to deliver them to his patron. Was Ledger not pleased?

Ledger embraced Mamani gratefully. After all his disappointments with the alpacas, the seeds must have seemed like the answer to his prayers. He carefully emptied them onto a sheet. There were forty pounds—in theory, more than enough to turn a healthy profit and revive his fortunes. He let the seeds sit in the sun for two hours every day for a month to make sure they were dry. Then he wrapped half of them in chinchilla skins, placed them in a box covered with hide, and, addressing the package to his brother George in London, shipped them to Britain via Cartagena and Panama. Next, he paid Mamani $500, and as an additional show of gratitude, gave him two mules, four donkeys, and a gun with ammunition. Finally, he told him that if he could procure another twenty pounds of seeds, he would pay him a further $600.

When Ledger sent the packet of seeds to George he had no idea of their true worth. All he knew was that the British, like the Dutch, had been willing to fund a series of expensive missions to South America in search of high-quality *calisayas*. As a subsequent letter to George makes clear, he was happy to put his trust in the British, confident that a grateful government would soon be beating a path to his door. "You might very greatly contribute towards obtaining for me a *recompense adequate to the services rendered*," he wrote to George in June 1865. "Sir William Hooker might bring the matter properly before the proper authorities and I feel that my services in this direction will meet with better return than my Alpaca toils."

Unfortunately for Ledger, his faith in British competence was misplaced. Sir William had just died, and when George arrived at the gates of Kew, he discovered that his son, Joseph Dalton Hooker, who had succeeded him, was ill and could not see him. The only person available was a superintendent who, not realizing the value of the seeds, declined to accept them and instead gave George a patronizing lecture on the cultivation of cinchona. Unknown to George, his arrival had coincided with the India Office's decision to end its relationship with Kew. By now, Spruce and Markham's seeds had reached India and were being raised in the government gardens at Ootacamund in Mysore. Kew, which had acted as a repository and nursery for the seeds and plants until they were ready to be shipped to the colonies, suddenly found itself superfluous. The only other person who could possibly have helped was Markham, but he had left for India.

Not knowing what to do, George began touting the seeds around London. His first stop was the Exotic Nursery in Chelsea, where he made a present of some seeds to the director, a Mr. Veitch (not realizing their value, Veitch innocently reported to George in November that the seeds were "germinating nicely"). Next, he wrote to the undersecretary for the colonies, offering to place the seeds at the government's disposal in the hope that when botanists had proved their value, his brother would receive a "just, fair and adequate" recompense from the government. George never received a reply. Instead, the undersecretary copied the letter to Joseph Hooker at Kew, asking him to take the matter up directly with Ledger.

In a panic, George offered the seeds to Professor Bentley of the Pharmaceutical Society, but he did not want them either and forwarded them to the Royal Botanical Society. George also gave some as a present to the Society of Arts, only for the society to send them to a Lady Hanbury. Finally, on Bentley's advice, George took the seeds to John Howard.

By now Howard was one of the leading quinine manufacturers in Britain and a preeminent authority on cinchona. More important, from Ledger's point of view, Howard was in regular contact with the Hookers at Kew and the Dutch, in particular with Franz Junghuhn and Karel Wessel Van Gorkum—respectively, the chemist and superintendent in charge of the cinchona plantations in Buitenzorg in Java.

George began by explaining how he had approached Kew and the undersecretary for the colonies, only to be passed from pillar to post. Howard agreed that it was pointless to try to offer the seeds around London when all the competent British authorities were abroad. But had he considered the Dutch?

On Howard's advice, George approached the Dutch consul general in London. The consul did not know what to make of the seeds, so he passed the inquiry on to Professor Miquel, a well-known Dutch botanist who had recently returned from making a collection of cinchonas in Java. Miquel was intrigued by George's story, and after confirming that the seeds came from Caupolicán—an area of Bolivia that Dutch collectors like Hasskarl had been unable to penetrate—he recommended that the consul should offer to buy one pound of the seeds. Given all the effort that Ledger and Mamani had gone to, the amount the Dutch offered was derisory—just 100 guilders (£20)—but at least it was a start.

GEORGE NOW HAD fourteen pounds of seeds left. Worried that they would soon spoil and be unsalable, he approached Howard again. Surely he must know someone else who was interested? Howard suggested a Mr. Money, the Anglo-Indian owner of extensive cinchona plantations in India, who was then vacationing in London. Money was not an expert and, at first, was unimpressed by George's explanation of how the seeds were the best in Bolivia and had been taken from an area that no European had ever ventured into. But after consulting Howard, and being assured that the seeds were viable, he agreed to buy them.

His offer was £50 for the lot, with a promise of a further £50 should ten thousand of the seeds germinate, take it or leave it. George took the money. Thus it was that the bulk of the seeds finally found their way via the back door to India—the place Ledger had originally wanted them to be cultivated—but not before they had also fallen into the hands of Britain's more commercially minded Dutch rivals. The British authorities' indifference to the seeds wounded Ledger deeply. He had risked his last savings on the venture and once again had nothing to show for it.

"I feel convinced in my own mind that no white man would or could succeed in getting such splendid seed as my faithful Manuel did," he wrote to Howard years later. "It is so clear that he got the true Calisaya Rojo. In fact, and as the good poor fellow repeatedly told me, he got seed from particularly fine old trees that we had together seen and sat under."

Gradually, the perceived slight would become a canker, eating away at his pride and fueling Ledger's sense of injustice. Again and again in his letters to Howard he dwells on what he sees as his harsh treatment by the British. Yet as far as the British authorities were concerned, he was just another South American businessman on the make. Like many an expatriate before and since, Ledger had made the classic mistake of believing that his devotion to the mother country would be reciprocated.

At this point a less stubborn man would have abandoned the cinchona scheme and concentrated on something that he knew was likely to turn a profit—such as trading alpaca hides again. Instead, Ledger sent Mamani back to the *calisaya* trees a second time. In so doing he sealed his friend's fate.

Ledger provides few details of the tragedy that resulted. All we know is that in 1869 he forwarded £200 to Mamani as a down payment on the next delivery of seeds. By now, Mamani had returned to his home in Coroico, a beautiful mountain village overlooking the Yungas—the continuation south of the same subtropical valleys where he had collected the first batch of seeds. We do not know whether on receiving the funds Mamani made a second expedition to Apolo or whether he chose this time to gather seeds from *calisayas* nearer his home. It seems likely, however, that it was the latter. As Ledger wrote in a long letter to Howard in 1877 from Argentina: "Poor Manuel is dead also; he was put in prison by [the] Corregidor of Coroico; beat so as to make him confess who the seed found on him was for."

According to Santiago, his father had been returning from the bark forests when he was arrested. Somehow, Coroico's chief of police had got wind of the fact that Mamani was collecting seeds to smuggle across the border, thus threatening Bolivia's valuable monopoly. He immediately had Mamani thrown in prison, where he beat him to try to get him to reveal whom he was working for. But Mamani stubbornly refused to betray Ledger. The chief of police withheld food and water for two weeks and beat Mamani again, but still he refused to reveal Ledger's part in the scheme. Finally, after twenty days the police chief released Mamani—as Ledger puts it—"beaten and half-starved."

The corregidor had taken everything Mamani owned—his donkeys, his blankets, and in all likelihood, his striped poncho and *mochila*, too. According to Santiago his father was barely able to stand up, and although the family did everything they could to nurse him back to health, his wounds were too serious and within a matter of days he was dead.

Ledger does not say whether he felt any guilt on hearing the melancholy fate of his friend. Santiago had ridden especially to Tacna to tell him the news and to account for the monies Ledger had advanced. According to Santiago, his father had been approached many times by other foreigners anxious to obtain seeds but had always refused, so it is probable his activities were common knowledge. But Ledger had also been fully aware of the risks, and the fact remained that had he not asked his friend to venture into the forests a second time, he would not have died. Incredibly, in spite of everything, Santiago himself now offered to return to the *calisaya* forests. This time Ledger wisely refused.

"Seeing the dangers to which these poor fellows were exposed to," Ledger wrote Howard, "having no direct application for any, feeling also that I had been miserably recompensed for what I had previously done, I told him to collect no more on my account or for any one else."

IT IS TYPICAL of Ledger that in lamenting Mamani's death he also cannot help reminding Howard of his own losses. His self-pity oozes from every line. But the more interesting question is what possessed

Mamani to take such risks in the first place. Why return to the *calisaya* patch three years in a row and then walk a thousand miles over the freezing cordillera? Was it simply for the money, or was it, as Ledger and Duran-Reynals would have us believe, because of his sense of duty and devotion to his master? Or did Mamani have other motivations we can only guess at?

Ledger once remarked on how whenever Mamani came upon a tree felled by a *cascarillero* he would drop to his knees and say a prayer for "*los buenos padres.*" According to Ledger, it was the Jesuit fathers who had impressed upon Mamani the importance of always planting five trees in place of every one felled in order that the forests should not become denuded. Perhaps the tree, not Ledger, was the real source of his devotion. After all, Mamani had seen with his own eyes how the *cascarilleros* were gradually destroying the *calisayas* to the point where supplies of Bolivian quinine would soon be exhausted. Could it be that, like Markham, he wanted to ensure that the valuable anti-malarial would not be lost to the world?

We will never know. Mamani is the point where the facts run out and the trail ends. Like the faded photograph from the Royal Pharmaceutical Society, he seems destined to remain forever slightly out of focus.

Whatever spin Ledger subsequently put on his role in the affair, however, it is clear he was motivated first and foremost not by philanthropy but by money. After selling the seeds to the Dutch and the British, he succeeded in placing the remaining twenty pounds, £100 worth, with the government of Queensland. But the funds were nowhere near enough to compensate him for his outlay. In all, he had spent £800 on procuring the *calisaya* seeds and had received just £170 in return. Yet any profits from the cultivation of the seeds would not accrue to him but to the Dutch or the aptly named Money.

In fact, Mamani's seeds would prove far more valuable than Ledger or anyone else could have imagined—and ironically it was the Dutch, not the English, who reaped the rewards.

9

A Tale of Two Seeds

He lurks among the reeds, beside the marsh,
Red oleanders twisted in his hair
His eyes are haggard and his lips are harsh
Upon his breast the bones show gaunt and bare

The green and stagnant waters lick his feet,
And from their filmy, iridescent scum
Clouds of mosquitoes, gauzy in the heat,
Rise with His gifts: Death and Delirium

— *"Malaria,"*
a lyric of India

IN 1864, FLORENCE NIGHTINGALE gave a prescient talk to the National Association for the Promotion of Social Science, in Edinburgh. The title of her address was "How People May Live and Not Die in India," and her theme was the urgent need for public health measures to combat the twin ravages of cholera and malaria then sweeping the Asian subcontinent. By the 1860s as many as two million people a year were dying of malaria in India, but as Nightingale pointed out, infections were probably running at "fifty times" that rate. "The drain upon human life and happiness of *fever* in India is literally untold," she began, but "the mortality from malaria in India is a mere trifle compared with the ravage fever commits in sapping the strength and vigour of the country, in making the young old, the healthy infirm for life, the industrious helpless invalids, the rich poor, the thriving country a waste."[1]

The malarious state of the country was only partly the fault of the mosquito. Thanks to the collapse of India's cotton-spinning industry and the East India Company's punitive land-tax policies, millions of workers had been displaced from the cities to the countryside, where they found themselves little better than serfs. Forced to pay up to half their crop to the tax collector, they sank deeper and deeper into debt and began to neglect essential public works—such as irrigation and canal building. The result was a huge increase in waterborne diseases such as cholera, dysentery, and, of course, malaria.

Ever since the Indian Mutiny had exposed the appalling conditions of British troops in Bengal, Nightingale had been campaigning tirelessly for improved irrigation and sanitation. Moreover, she had quickly realized that in India—unlike in the Crimea—malaria and cholera were not simply problems confined to barracks. From the Punjab to Madras, and Burma to the Western Ghats, malaria had laid waste to large swaths of countryside, and even where the fevers had not brought, in the words of the popular lyric, "death and delirium," they had so weakened the rural labor force that drainage works had gone neglected. Nightingale concluded that the only way to protect the army was to institute sanitary measures throughout the country, but this would take years and the army had to plan for a revolt next month. One idea was to station British troops on high ground, well away from the unhealthy "miasmas" associated with low-lying swamps and jungles. But many of the hill stations in India were also malarious, and if another mutiny broke out they would be isolated. Now, thanks to Markham and his "fellow labourers," another solution was at hand.

The story of quinine after Markham's departure from South America is essentially a tale of two seeds: Spruce's *Cinchona succirubra* and Mamani's *C. calisaya*. It is a story that pits Dutch nurserymen and chemists against British gardeners and bureaucrats, or, to put it another way, hardheaded realists against well-meaning but muddled "amateurs." But it is also the story of how the liberal instincts of health reformers like Nightingale combined with the enlightened self-interest of the British ruling classes in India. And running through the story is a question that bedevils the debate over the development of antimalarials to this day: Can new malaria treatments be secured through the pursuit of profit alone, or do they necessarily entail an element of philanthropy?

Interestingly, when the British and the Dutch first set out to raise cinchona in their colonies, they assumed that philanthropy would be enough. Just as Markham had announced that his aim was to make a cheap cinchona febrifuge available to "every pansari's shop" in India, so in 1854, Rochussen, the governor-general of the Netherlands Indies, declared that the Dutch effort was "no common speculation. The eyes of Europe are fixed upon this enterprise which interests it in the highest degree; for it is, so to speak, to procure for the sick, the means of life, and to procure it at a cost which will not exceed his means."

In fact, neither Britain's nor Holland's policy was quite so clear-cut. The priority of civil servants at the India Office was the health of the British army in India and the well-being of the East India Company's white employees. If the cinchona missions, by securing increased supplies of cheap quinine, also brought the drug within the reach of the brown masses, then so much the better. Similarly, while in Java the Dutch government sponsored the initial experiments with cinchona—effectively underwriting the costs of developing a chemically lucrative strain—the decision whether to farm the tree was in the hands of private planters. Unless they could be shown, or given guarantees, that cinchona was more profitable than coffee or sugar, philanthropy alone was never going to induce them to grow it.

The problem was that transporting seeds and plants to the colonies and developing the plantations was a far more complex horticultural and pharmacological task than anyone had anticipated. Indeed, for Markham the difficulties began the moment he dispatched his collections from the coast of Peru.

Before setting off for South America he had taken every precaution to ensure that his plants would survive the voyage to India. Mindful of Hasskarl's disastrous experience in 1854, he visited the Horticultural

Society expressly to consult the tea collector Robert Fortune about the best design of Wardian cases. As a result of those discussions, Markham commissioned thirty state-of-the-art cases, complete with battens for keeping the soil compacted and the plants separated into neat rows, and putty to ensure the glass seals were as airtight as possible. On arriving in South America, Markham deposited fifteen of the cases at Guayaquil ready to receive Spruce's collection, and brought the remaining fifteen with him to Islay. In addition, at Lima, he had a further six cases built to receive Pritchett's gray barks from Huánuco.

The only thing Markham didn't do was fit the cases with a false bottom. This innovation, designed to improve air circulation and drainage around the roots, would come later. In Markham's case it would probably have made little difference.

On leaving Peru, Markham's Wardian cases contained more than 450 seedlings. The original plan had been to send them directly to India by Royal Navy frigate, but the promised ship never arrived, and instead Markham was forced to book passage on a commercial steamer via Panama. By the time he and John Weir docked at Southampton in August 1860, only 270 of his plants were still alive. The sensible course of action would have been to nurse the remaining plants at Kew and send them on to India at the end of the summer. But for some reason Markham decided to proceed straight to Bombay and the Nilgiri Hills, the site in southwest India he had chosen for the first experimental plantation.

The result was disaster. In crossing Egypt, the cases were knocked about and one of the most valuable lost, dropped in the Red Sea, according to Markham, by "truculent Arabs." Next, the ship's engines failed, and for twenty-four hours the ship was stalled in the sweltering heat of the Arabian summer, with temperatures rising as high as 107 degrees on deck. Then, on reaching Bombay, he learned that the steamer he had been planning to connect with had already left for Calicut, forcing him to remain in the bay for several days, where the plants were exposed to the "very pernicious climate." It was not until mid-October—three and a half months after leaving Peru—that the plants finally reached the government plantations at Ootacamund, sixty miles south of Mysore. Markham had selected the Nilgiri Hills because of their resemblance to Caravaya, the province in Peru where he had gathered the plants. At Ootacamund he could also call on the skill of an expert horticulturist, William McIvor, the superintendent of the government gardens. Unfortunately, by the time the plants reached McIvor it was already too late. He planted them out, but the delays had been fatal, and by December all of Markham's *calisayas* were dead.

After all he had been through—the grueling descent to the Tambopata River, the flight across the freezing cordillera to Arequipa, and his outwitting of Peruvian customs—it must have been a bitter blow. But in his subsequent account of the introduction of cinchona to India, Markham makes no comment on the loss. One possible reason is that by 1880, when he came to publish *Peruvian Bark*, it was already apparent that the most valuable trees were not his but Spruce's and Mamani's. Under the circumstances, he probably considered there was little point in harping on his misfortune.

There may have been another reason. The year after he left South America, Robert Cross had followed exactly the same route to India without any problems. Cross set sail from Guayaquil with Spruce's *Cinchona succirubra* in January 1861, landed at Southampton in February, and then proceeded across the Red Sea, arriving at Ootacamund in April with 463 healthy plants. Markham later attributed Cross's success to the fact that he had crossed Egypt in winter when the climate was cooler. But the success also drew attention to the fact that Markham was not a trained botanist, and it was not long before aspersions were being cast on the methods being used by Markham and McIvor to propagate cinchona in India.

Their principal critic was Thomas Anderson, a young botanist who had been dispatched to India by Sir William Hooker in 1862 with *Cinchona succirubra* and *C. calisaya* seeds carefully raised in Kew's nurseries. When Anderson, who had also visited Java, saw how Markham and McIvor were running the government plantations, he was not impressed. In a report to the India Office, he implied that Markham had played down the failure of his first batch of plants and had capitalized on the success that should have been due to Spruce and Cross. He also argued that the method of cultivation in the Nilgiri Hills was defective.

Ironically, what saved Markham and McIvor from ridicule was Spruce's *Cinchona succirubra.* The tree did so well at Nilgiri it was as if nature had designed it with India in mind. In accordance with Spruce's instructions, McIvor planted the saplings on an exposed site at Nedivattam, in the lower Nilgiri Hills, where they were guaranteed plenty of rain and sun. Meanwhile, at another site higher up the same mountain range, he began raising plants and seeds of *Cinchona officinalis* brought by Cross from Loja. By 1866 he had more than 244,000 trees planted out. To Markham, the transformation seemed miraculous. "The old jungle had disappeared," he wrote, "and in its place were the rows of Peruvian bark trees with their graceful and beautiful foliage."

In describing the success of the *Cinchona succirubra* trees, however, Markham glossed over the fact that by now the *C. calisaya* seeds he had collected in Peru had also arrived in India. Owing to the lack of rainfall and differences in the soil and elevation, however, they did not fare nearly as well. In fact, the rootstock was soon attacked by parasites, and McIvor had great difficulty getting any to grow at all. Under the circumstances, he had little choice but to concentrate on Spruce's and Cross's species, with the result that by 1878 there were some 305,000 *Cinchona officinalis* trees, 261,000 *C. succirubra*, and just 552 of Markham's *C. calisayas* growing in the government plantations. As far as the India Office was concerned, Markham was on the right track and would soon be in a position to manufacture sulphate of quinine at a greatly reduced cost.

However, the faith of Markham's paymasters in London was misplaced: he would never be in a position to manufacture quinine cheaply—at least not as cheaply as the Dutch—for the simple reason that Mamani's seeds were far more productive than Spruce's. Fortunately for Markham, however, the red bark was rich in other alkaloids that he now proved were just as effective at clearing the malaria parasites and that would make his dream of a cheap febrifuge a reality.

IT IS INTERESTING to note how close the Dutch plantations were to collapse when the one-pound bag of seeds that George Ledger had sold to the Dutch consul in London in 1865 reached Java. For more than ten years, Dutch botanists and chemists had been struggling to develop a high-quinine-bearing strain at the government nurseries at Tjinjiroean. Thanks to Hasskarl, however, the experiment had gotten off to a disastrous start. Not only had he chosen the wrong mountain range, but the species he brought back from Peru—*Cinchona pahudiana*—contained hardly any quinine at all. Under the direction of Hasskarl's successor, Franz Junghuhn, the Dutch decided to concentrate instead on plants raised from the original *C. calisaya* seeds collected by the French botanist Hughes Weddell in Bolivia in 1850. Then, in 1862, they began supplementing Weddell's *C. calisaya* with *C. succirubra* and *C. officinalis* trees that McIvor and Markham had forwarded in a spirit of cooperation from the British nurseries in India.

Interestingly, it was the Dutch who first established that *C. succirubra* was rich chiefly in cinchonidine, cinchonine, and quinidine. Indeed, in one assay the Dutch chemist de Vrij found that Spruce's bark contained as much as 8 percent of the febrifugal alkaloids. But since Java's principal aim was quinine production, the results were of little interest to the Dutch planters. Soon they were mocking the government cinchona estates as "the Governor's expensive hobby," and by the time Mamani's seeds arrived, many were on the point of abandoning cinchona cultivation altogether.

At first there was little indication that Mamani's seeds would be any different from all the other expensive failures. When the new superintendent of the gardens, Karel Wessel Van Gorkum, opened the small package from the consul in London, it gave off a strong smell of ammonia, and he feared the seeds had spoiled. Nevertheless, he managed to get twenty thousand to germinate and within a year he had twelve thousand healthy young seedlings. However, at this stage Van Gorkum writes that he had "no reason to suppose that we had come into possession of an especially valuable description of cinchona; on the contrary, we thought we were dealing with the identical *calisaya* we already possessed."

Looking back, the Dutch could easily have compromised their own experiment. From a geneticist's viewpoint Java was one of the worst places for Mamani's seedlings. By 1865, the Dutch colony was full of flowering cinchona trees of dubious commercial and genetic value. Because of cinchona's propensity to hybridization, once Mamani's trees flowered, any one of the other species in Java could have cross-pollinated with them, contaminating the strain and diluting its genetic purity before the bark could be tested for quinine. As Mamani's *C. cali-*

saya had never been cultivated before and Van Gorkum had no way of knowing when they might flower, this was no slight risk.

Fortunately for the Dutch, a brilliant young chemist had just arrived on the scene. J. C. Bernelot Moens had been sent to Java with express instructions to determine the alkaloid content of the trees in the government plantations. Unlike the British—who had to send their barks all the way to John Howard in Tottenham for analysis—Moens had all the equipment he needed installed on site. Beginning in 1872, he very carefully took strips of bark from the trees for testing. He had no great expectations. He had already tested a number of other cinchonas on the island, broadly confirming de Vrij's findings that they contained disappointingly small amounts of quinine. Imagine his surprise then when the results of his first tests on Mamani's *C. calisaya* came back. The young trees contained 8 percent quinine—far more than any other species of similar age. Two years later the quinine content in the maturing trees had risen to 11 percent, and by 1876 it had reached 13 percent.

It was an unheard-of amount. In contrast, Moens found that the *C. calisaya* trees raised from Weddell's seed contained a maximum of only 2 percent quinine, while the variety that had survived from Hasskarl's ill-fated expedition contained even less. Indeed, the only species that bore comparison was Cross's *C. officinalis*. In 1873, de Vrij had visited Ootacamund and discovered that some of the offspring of Cross's Loja collection contained up to 9 percent quinine. But later tests in Java confirmed that the alkaloid content of the barks varied widely depending on the age and variety, and that in the case of some trees it was as low as 1 percent.

It was rapidly becoming clear to Van Gorkum that the one-pound bag they had purchased from George Ledger contained a very unusual variety of *C. calisaya* indeed. But it was only when he sent 260 kg to be auctioned at Amsterdam in 1877 that he realized just how unusual: it fetched 17.5 florins a kilo, double its reserve price; in contrast, a kilo of *C. succirubra* bark fetched only 3.4 florins. "The reputation of this description of cinchona was thus assured," writes Van Gorkum, "and has since splendidly maintained itself at each auction."

In retrospect it is surprising that the news did not spark an orgy of speculation on a par with the tulip craze of 1630. Then riots had threatened to break out in Amsterdam after a single bulb sold for $10,000. Mamani's bark had not fetched quite so spectacular a sum, but in view of quinine's medicinal importance—after more than three hundred years it was still the only drug that cured malaria—it represented a far better long-term investment. The problem was, and this is the reason the Dutch reaction was more sober than in 1630, the investment *had to be* long-term. It takes at least fifteen years for a *C. calisaya* tree to mature to the point where its quinine content is at a peak and its bark can be harvested. To ensure a constant supply of quinine, the Dutch planters would have to plant new seedlings every year so they would always have new trees to replace ones that had been stripped the previous season. That entailed not only sufficient land to run several plantations in parallel, but investors with pockets deep enough to fund the plantations for fifteen years before they saw a return on their money. The problem was particularly acute in Java, as the island was already overcrowded and the Dutch had to retain the older, less economically viable species of cinchona until Mamani's trees came onstream.

Perhaps the Dutch had learned the lesson of the tulip craze, or perhaps after so many disappointments with cinchona in the past they were more hardheaded. Van Gorkum now drew up a strict plan that contrasted sharply with the haphazard cultivation methods being pursued by McIvor in the Indian plantations. In the future, Java would concentrate on just three species: Mamani's *C. calisaya* and Cross's *C. officinalis* for quinine, and Spruce's *C. succirubra* for the other alkaloids. To ensure that there was no contamination while inferior species were being phased out, Van Gorkum cleared a small area of virgin forest for Mamani's *C. calisaya* away from the other trees and, as an additional precaution, had thousands of worthless varieties surrounding the plantation deflowered every year. It is no exaggeration to say that these few simple precautions saved the entire Dutch cinchona effort. In 1879, when the first trees were ready for harvest, most quinine bark still came from South America. However, by the 1930s, Java was providing 97 percent of the world's supply.

The question is why the British didn't achieve similar success. After all, through the Indian plantation owner Money, they had acquired far more of Mamani's seeds than the Dutch. Moreover, whereas in

1865 the Dutch government was under fire for squandering money, the Indian government had spent hardly anything, and Markham and McIvor—after weathering the initial criticism—enjoyed pretty much free rein.

The short answer is that Markham's primary motivation was philanthropic, and as far as he and McIvor were concerned, they had already found a tree perfectly adapted to India's needs—Spruce's *C. succirubra.*

By 1872, CINCHONA CULTIVATION had spread to every corner of India. Following the success of the government plantations in Nilgiri, private plantations had sprung up in the coffee-growing districts of the Coorg, Travancore, and Tinnevelly, as well as in Darjeeling in British Sikkim, Burma, and Ceylon. All told, India now had more than four million cinchona trees, and there was little doubt about the planters' favorite. Cross's *C. officinalis* might be the choice at higher elevations, but on the lower slopes *C. succirubra* was indubitably king. In Ceylon alone there were nearly one million red bark trees, prompting one visitor from the Pharmaceutical Society to comment:

> *Succirubra* is the variety in favour, the variety in demand, and the variety that thrives. It may be found growing luxuriantly everywhere, in the mountain provinces, on fertile estates, and on the poorest patena soils . . . So familiar indeed has it become to the eye of both colonist and native, that it is looked upon as indigenous; the latter having now the fullest confidence in it as a remedy, and not hesitating to help himself to a strip of bark from the nearest tree when occasion necessitates it.

But the preponderance of *C. succirubra* was not dictated simply by horticultural considerations: it was also the result of policy choices forced on the plantations early on by Markham and McIvor. By making his aim a cheap febrifuge for the masses, Markham had set out his stall. Unlike the Dutch, he did not have the luxury of waiting for the government nurseries to produce a strain that was so rich in quinine its commercial advantages would be apparent to all. Hence, his memorandum to the Revenue and Judicial Committee in 1865 starkly posed the question: "Did the Government undertake Cinchona cultivation in order that the use of quinine, in some form or another, might be extended to the people of India, now entirely debarred from its use, or did they undertake it as a mere speculation?" If the latter, then the bark should be sent back to England and sold for the manufacture of quinine. But this would mean that the Indian masses would effectively be deprived of the use of the febrifuge. "The object of the undertaking will be frustrated, and the Government will have entered upon a speculation which will no doubt pay well, but which is entirely foreign to its functions."

Clearly, Markham was worried about confusion within the Council of India regarding the purpose of the missions. By 1865, he was back in London serving as a clerk to the Board of Control's Secret and Confidential Department and was well placed to observe the council's innermost thoughts. There are hints in his memos that he may also have been seeking a reason to return to India. Interestingly, it was about this time that he began arguing for the preparation of a cheap febrifuge from all the alkaloids which could be supplied "from thousands of village dispensaries and [which] would be within the reach of every *ryot* [peasant] in the land." What better excuse to return to India than the need to oversee experimentation on the bark?

The other factor was McIvor. He had already decided there was little point in trying to cultivate the quinine-rich *C. calisaya* in India because of the problems the trees had acclimatizing. Instead, he was concentrating all his efforts on *C. succirubra* and *C. officinalis*. Unlike the Dutch in Java, who had taken pains to keep different varieties of the same species separate, McIvor was freely allowing hybridization in the hope that a cross between *C. succirubra* and *C. officinalis* or *C. calisaya* would spontaneously produce a hybrid higher in quinine than either. He also experimented with "mossing" and coppicing techniques to boost the alkaloid content of the bark. John Eliot Howard, who was intimately acquainted with both the Indian and Javanese plantations, viewed McIvor's approach with scorn. "In Java they are wisely thinning out the inferior trees and allowing the best to develop themselves. In British India, on the contrary, they are acting in a way that is as consistent with common sense as to believe that by fusing together a half-crown and a penny, one could produce a sovereign."

As far as Howard was concerned, the sovereigns were Mamani's seeds, which he had dubbed *Calisaya ledgeriana* in recognition of the role Ledger had played in forwarding them to Europe. Indeed, Howard was so impressed with the Dutch results that in his monolithic *Quinology of the East Indian Plantations*, he described *Calisaya ledgeriana* as the "legitimate descendant of the finest sort of the Bolivian forests," and "the best, by far, of all."

As the world's leading authority on cinchona and the pharmacology of the bark, Howard's opinion ought to have counted for something. But in the case of India, it seems, Markham and McIvor were a law to themselves.

Money had been reluctant to buy George Ledger's seeds, so it is not surprising to discover that soon after returning to India, he had swapped them for a like amount of *C. succirubra*. But when the seeds reached McIvor, all his attention was concentrated on *succirubra* and *officinalis*. Although he germinated the *ledgeriana*, planting some in Ootacamund and sending the rest to Sikkim, where he believed the climate would be more suitable, either through lack of interest or neglect the trees did not do well and were soon forgotten.[2]

Under any other circumstances, this neglect would have been criminal. Not only did India have one of the highest incidences of malaria in the world, but in 1872 a series of deadly epidemics swept through Bengal, underlining the need for a plentiful and cheap supply of quinine. Fortunately, by now Markham had established that the other alkaloids in *succirubra* were almost as good at curing fevers and that one of them—quinidine—was quinine's equal.

Ever since Louis XIV had revealed Robert Talbor's secret formula for the "English remedy" to the world, it had been known that a rough decoction of powdered bark was usually sufficient to cure malarial fevers. But following the initial studies on quinine, no one had thought to reexamine the other alkaloids in the bark to see if they might be equally efficacious. This was because by the 1870s manufacturers had perfected the extraction of quinine from cinchona bark in a form that was stable, easy to transport, and could be administered to patients in controlled doses. But while quinine hydrochloride—as the refined drug was known—was well within the means of wealthy Europeans, Markham feared that because the extraction process was laborious and expensive, the production costs would never be low enough to bring it within the reach of India's impoverished masses. First, the bark had to be mixed with alcohol and lime, heated, and filtered to obtain the alkaloid bases. Then those bases had to be treated with a series of precipitants—usually sulfuric or hydrochloric acid—to separate out the salts. Finally, to separate the quinine from the other alkaloids, the mixture had to be filtered a third time, treated with caustic soda, dissolved in alcohol again, and left to crystallize.

What, however, if a mixture of all the alkaloids was sufficient to cure malaria fevers? Then there would be no need to go to all the expense of extracting pure quinine, and the remedy would be affordable to everyone.

Beginning in 1865, Markham began lobbying for tests on the other alkaloids in the bark to see if Talbor's intuition had been correct. His argument that local circumstances in India might favor a compromise febrifuge soon struck a chord, and in 1866, Viscount Cranbourne, secretary of state for India, agreed to the assays, ordering the establishment of Cinchona Commissions in Calcutta, Madras, and Bombay, staffed by medical officers with the task of conducting the experiments.

By today's standards, the experiments were very rough-and-ready. Scientists had yet to identify the malaria plasmodium, so instead of microscopic examination of blood smears, the medics had to rely on their readings of patients' symptoms to determine whether or not the alkaloids had cured the disease. There were also no controls. Instead, each of the alkaloids was administered to a different set of patients and the results tabulated. Nevertheless, taken at face value, the results were astonishing. In the case of the Madras Commission, 846 patients were treated with quinine, 664 with quinidine, 559 with cinchonine, and 403 with cinchonidine. In each group, the alkaloids separately cured the vast majority of fevers. Cinchonine fared the worst with thirteen failures, while quinine came second with six, and cinchonidine and quinidine tied with four. However, the quinine and quinidine failure rates had to be set against the fact that they had been administered to a much larger number of patients.

The commission's conclusion was that quinidine and quinine possessed "equal febrifugal power," while cinchonidine was "only slightly less efficacious." Even cinchonine, "though considerably inferior to the other alkaloids [was] notwithstanding a valuable remedial agent in fever."[3]

Not surprisingly, Markham seized on the results, arguing that the government's aim should now be a "mixture of all the alkaloids which would combine cheapness and efficacy in the highest degree." If anything, Howard was even more enthusiastic, arguing that all ordinary Indians needed to do was prepare a decoction from the bark—a process so simple "it may be carried on by every family that has a few trees planted at its door."

However, it was Markham's proposal that received approval. He was aided by the fact that although private planters were still pressing for a species of cinchona that could compete with *C. ledgeriana* in the international market, the India Office was rapidly losing patience with their demands.

In 1866, for instance, the Madras government had recruited an eminent chemist, John Broughton, to investigate the manufacture of a cheap febrifuge that would retail at no more than one rupee an ounce. Broughton succeeded, establishing a factory in Nedivattam that produced a febrifuge he called amorphous quinine at cost price. But the experiment was scuttled by government accountants who wanted to see a greater return on the sale of the bark from the Madras plantations.

Markham was furious, blaming the accountants for Broughton's subsequent resignation, and arguing that their only object was "to obtain harvests of bark to sell at profit in the London market." Markham soon won the backing of a Mr. Hume, a secretary in the India government's agriculture department, who had seen at first hand how malaria had undermined the health of the rural labor force. So when the Sikkim government appointed another chemist, C. H. Wood, to organize the manufacture of a cheap febrifuge, his argument that the government would not want to sell its alkaloids so cheaply as to undermine private planters in "the bazaars" met with an immediate and furious response. "The question for Wood to consider," fumed Cranbourne, "is not what kind of bark fetches the highest price in the London market . . . but to grow the bark yielding the largest percentage of the febrifuge alkaloids generally . . . It should be understood that the object of Cinchona cultivation . . . is not a commercial speculation."

Whether or not he realized it, Cranbourne's sentiments exactly echoed those of the governor-general of the Netherlands Indies twenty years before. Wood got the message and began experimenting with quinetum—an extract of the alkaloids, invented by de Vrij in 1872, produced by dissolving the bark in hydrochloric acid and then repeatedly filtering and washing the resulting powder. Although the process was unsophisticated, it had the advantage of being cheap, with the result that quinetum retailed for one rupee an ounce without loss to the Sikkim government.

By 1880, Markham could justifiably look back on cinchona cultivation in India with pride. India was now self-sufficient in bark. Not only that, but the barks from the Nilgiri plantations were commanding higher prices in the London market than those from South America. Moreover, although the Indian plantations would never be as lucrative as the Javanese, the whole cost of introducing and cultivating the tree in India had been repaid with profit. Markham could not resist pointing out that even as a "mere commercial speculation," his enterprise had been a success. The Dutch had not been idle either. Thanks to their careful nurturing of *Cinchona ledgeriana*, they were also about to bring quinine within reach of the masses, and ironically, they would do it by using Spruce's trees in a way that had yet to occur to the British.

ALTHOUGH MOENS HAD ESTABLISHED that *C. ledgeriana* was the species richest in quinine, not all the offspring of Mamani's seeds contained equal levels of the alkaloid. Indeed, some of the trees contained as little as 4 percent, suggesting that they had not all been taken from the same area of Caupolicán or that the strain was itself a hybrid of others. Moens and Van Gorkum had already taken steps to ensure that *C. ledgeriana* did not hybridize with other species of cinchona in Java. Now they began refining the selection process. First, two trees with a high percentage of quinine were chosen. Next, the best seeds were selected and carefully nurtured to maturity. Then, once the trees had flowered, the descendants possessing the highest percentage of quinine were reselected. The result was a second-generation strain with 25 percent more quinine than the first.

There was just one problem: the new, quinine-rich strains of *C. ledgeriana* were even more difficult to grow than their predecessors. The rootstock was prone to attack by fungus and only thrived in virgin soils, which were in short supply in Java. By contrast, *C. succirubra* had proved exceptionally robust and, as long as it was not overly shaded, flourished everywhere. The solution was obvious: in the future, shoots of young *C. ledgeriana* would be grafted onto *C. succirubra*. Not only would this give them a greater chance of survival, but the grafting would ensure that the bark quality of the original selected tree was perpetuated from one generation to the next—something that could not be guaranteed with propagation by seeds alone.

The results were spectacular. In 1873, when Moens and Van Gorkum began their experiment, there were just ten thousand *C. ledgeriana* in Java. By 1880, there were nearly half a million, plus more than six hundred thousand *C. officinalis* and *C. succirubra*.

The Dutch success had a dramatic effect on world quinine prices. In 1880, the majority of the bark traded on the London market still came from South America, and though the price had begun to slide, quinine sulphate traded at nearly 500 French gold francs a kilo. By 1890, as the Javanese bark started to come onstream, the price had fallen to 60 francs a kilo, and seven years later it was down to about 30 francs.

The collapse in the price of quinine was excellent news for the millions of poor infected with malaria each year but a disaster for private planters and the native South American bark industry. One after another, cinchona plantations in Darjeeling, Madras, and Ceylon had to be dismantled and sold off for the "more successful culture of tea." Belatedly, the Indian planters had tried to follow the Dutch example, replacing Spruce's red bark with grafts of *C. ledgeriana* grown from the original seeds purchased by Money. But they failed to produce the same quinine-rich strain, probably because the trees had already hybridized as a result of earlier neglect.

However, it was in South America—and Mamani's native Bolivia in particular—that the collapse in the price of quinine was felt most keenly. In 1851, the *calisaya* forests of Caupolicán and the Yungas produced three million pounds of bark. Thanks to the Bolivian government's monopoly, the trade brought $142,000 a year into the public treasury, accounting for 15 percent of the republic's total tax revenues. The money from quinine not only funded essential public works, such as roads, but also financed the construction of La Paz Cathedral, one of the city's finest buildings. By 1880, however, Bolivia, Peru, and Ecuador were exporting just one million pounds of bark combined. The effects of the quinine crash were felt most brutally in Sorata, a spectacular mountain retreat nestled below the summits of Illampu and Anchohuma, northeast of La Paz. In 1875 a group of German plantation owners had begun building a spectacular trail from Sorata down to Mapiri, a steamy jungle town on the banks of the Mapiri River. The idea was to halve the time it took to bring cinchona bark up the mountain to Sorata, from where it could be sent directly to La Paz by road and over the cordillera to the coast of Peru. The ascents were brutal, but worst of all was the *cuesta de amarga*, the "slope of bitterness." It was aptly named. By the time the Mapiri trail was opened in 1879, the price of quinine had already begun to fall, and by the turn of the century the plantation owners had been forced to uproot most of the cinchona trees and replace them with rubber.

Nevertheless, the success of the British and Dutch plantations should have been a source of great satisfaction to the two men who had made them possible. Both Spruce and Ledger had seen how the reckless stripping of the bark forests in South America had threatened to make the best species of cinchona extinct, and to a greater or lesser degree both shared the conviction that only by exporting the seeds and plants to the colonies could the survival of this "needful vegetable substance," as Spruce had put it, be secured for humanity. Ironically, no sooner were the cinchona plantations thriving than they, too, came to view them with bitterness. The focus of Spruce's resentment was

Markham and the India Office, while for Ledger it was primarily the Dutch. But in both cases their dissatisfaction came down to the same thing: money, and what they regarded as inadequate public recognition for their achievements.

After Spruce sent his collection to India, his health deteriorated rapidly. Whether it was a return of the mysterious paralysis that had visited him in Ambato, malaria, or some other tropical disease contracted during his exertions in the bark forests and the descent to Guayaquil is unclear. With Spruce's luck it could have been all three. He became feverish again and was in such pain that he could barely sit upright to complete his report to the India Office. For five months—from May to September 1861—Spruce lay in his hammock, writing a little each day and stretching his legs when his paralysis abated sufficiently to allow him to walk. Then in October he suffered another blow. After waving goodbye to Robert Cross, Spruce had deposited his entire life savings—£700, including the £360 he had received from the India Office—in a bank, Gutierrez & Company, in Guayaquil. But Gutierrez & Company was now declared bankrupt, and all his savings were gone. In desperation he appealed to the Foreign Office for some kind of consular position in South America that would pay him a small income and allow him to continue botanizing. But no offer was forthcoming and he became more and more despondent as he envisioned himself dying of fever and poverty in the marshes of Guayaquil, never able to return to England and the company of his scientific colleagues. In a letter to Daniel Hanbury in October 1861, he complained: "Had I worked as long in the East as I have done in the West, I might have reasonably calculated on a small pension when I was disabled from working; but I do not suppose there is any hope of such a thing in my case."

If Spruce's condition had improved, as it did on previous occasions, his spirits might have lifted. But this time his paralysis worsened. He was able to make only two further short expeditions, one to the plain of Guayaquil, the other to the coast, and by the end of 1862 he was so disabled that further collecting was out of the question. Writing again to Hanbury he gave full vent to his bitterness:

> I had never calculated on losing the use of my limbs, and yet nothing was more likely to happen, if the sort of life I led be considered. When after loss of health came wreck of fortune, simple though my wants be and modest as were my aspirations, I felt for a time completely prostrated. The fact is I have been far too constant to botany—several times in the course of my travels I might have taken to some occupation far more lucrative—and I have met many men, who beginning without a cent have made far more money in 2 or 3 years than I in 13, and that without being exposed to thunderstorms, and pelting rain—sitting in a canoe up to the knees in water—eating of bad and scant food once a day—getting no sleep at night from the attacks of venomous insects—to say nothing of the certainty of having every now and then to look death in the face, as I have done.

In his next breath Spruce asks Hanbury to "excuse these personal details," but clearly his sense of injustice was mounting. "Too late I have found that there is no sympathy for me where most I might have expected it," he declares.

In 1863 Spruce traveled to Paita, in northern Peru, in the hope that the drier desert climate would prove more conducive to his health. He succeeded in making excursions to Piura and Chira and drew up a report, at Markham's request, on the Peruvian system of cotton cultivation. But the miraculous recovery he had been hoping for never came, and by 1864 he had run out of options. His health broken by his adventures in the Amazon and Andes, and with no funds to speak of, he now faced the prospect he had feared all along—a penurious return to England. Six years earlier he had written to Bentham: "Poverty is such a positive crime in England that to be there without money or lucrative employment is a contingency not to be reflected on without dread." Now, as he bid goodbye to South America—his home for the last fifteen years—the dreaded prospect was staring him in the face. After all his exertions in the cause of botany, the least he could have expected was a small welcoming committee. In the course of his travels he had collected seven thousand flowering plants and ferns, as well as numerous maps and Indian vocabularies. According to Bentham, his researches were the most important "since the days of Humboldt." Yet when he disembarked at Southampton on May 28, 1864, there was no one to meet him. Instead, he went straight to Kew, where, between spasms of pain, he threw himself into work, sorting out his massive collection. His labors would eventually result in the publication, in 1885, of his six-hundred-page *Hepaticae of the Amazon and of the Andes of Peru and Ecuador*, which described seven hundred species and varieties of plants, four hundred of which were new to botany.

It was a monumental effort, the more so because—although he was now back among friends—his health continued to deteriorate. Although the reclamation of marshland in Essex, Kent, and Norfolk had gradually reduced the endemicity and extent of malaria in southern and eastern England, every now and again "ague" epidemics would flare up without warning. As late as 1887, hospital admissions for intermittent fever throughout Britain were running at 112,000 cases, and in London the boroughs of Lambeth and Westminster were particularly badly affected. Whether Spruce had been passing through these areas, or was just unlucky, we will never know, but shortly before Christmas 1864 he suddenly developed a fever and began to shiver. A doctor was called to his bedside at Kew and immediately diagnosed ague. Oddly, in view of the fact that this was clearly not Spruce's first experience of malaria, he wrote to Hanbury that in all his time in South America he had never suffered an "intermittent fever," although "I have had my full share of *remittent* and *continued* fever" (Spruce's italics). Interestingly, he included in the latter categories the fever that had nearly killed him on the Orinoco, suggesting that he only considered intermittent fevers to be true malaria. As before, he took quinine—ample supplies of which he had brought back with him from Peru—and he was cured.

His paralysis, however, continued to defy treatment. On Hanbury's advice he consulted a number of prominent specialists in London, but the symptoms continued, so much so that, according to Alfred Wallace, "for months or even years together, he was unable to sit up at a table to write or to use a microscope, and could never do so for more than a few minutes at a time with intervals of rest on a couch."

It was a country doctor—a Dr. Hartley of Malton—who eventually identified the cause as a "stricture of the rectum." Spruce had signaled the area as the "seat" of his disease for years but said that the specialists had preferred to ascribe his symptoms to hypochondria and never so much as thought of "passing a bougie into the rectum."[4] Now, following treatment with enemas and gentle opiates, the pain subsided and Spruce was once again able to work at his microscope and take short walks. But his condition had become chronic, and the discomfort remained with him until the end of his life.

By now, Spruce's influential friends had rallied to his cause and had set him up in a small cottage at Welburn, near Castle Howard, not far from the village where he had been born. Thanks to the intervention of George Howard, the Earl of Carlisle, his plight had also been brought to the attention of Lord Palmerston, the former prime minister, and in 1865 the government agreed to award him a small pension, of £50 a year. In view of the large revenues beginning to be generated by the Indian plantations, it was a miserly sum, and hardly enough to live on. Markham was appalled and immediately wrote to the Revenue Department, calling on the government to increase the pension to £100. The India Office refused, granting Spruce a onetime payment of £200 instead. Incensed, Markham wrote again, pointing out that Spruce was still owed £97 for his work in Ecuador and that more than any other collector he merited special recognition for his "great and valuable service" to India in gathering *Cinchona succirubra.* The India Office sanctioned the £97 payment, but it was not until 1877 that it reluctantly acceded to Markham's demand to grant Spruce a further £50 annual pension.

According to Markham, Spruce's modesty and "high sense of honour" prevented his making representations on his own behalf. But in a letter to Howard in 1869, Spruce privately gave expression to his anger, and interestingly, he appears to have blamed Markham. Howard had written to ask Spruce where he could obtain information about his role in introducing *C. succirubra* to India. Spruce replied that the information should have been included in the "Blue Book"—the parliamentary record of the correspondence relating to Markham's missions—but for some reason had been left out. Spruce pointed out that he had first proposed gathering cinchona in Ecuador in 1858—two years before Markham—but that his proposal had been held in abeyance by the Foreign Office until the Indian Mutiny had been "definitely quelled." Then, without mentioning Markham by name, Spruce continued:

> Afterwards, as you know, the direction of the enterprise was put into other hands and this is perhaps the reason why the whole of the above-mentioned correspondence was omitted from the report; thus robbing one of the merit (very slight, possibly) of having been the first person who, already practically acquainted with the country and the plants, proposed to procure seeds etc of the Peruvian Bark for forming plantations in the East Indies and other of our colonies, and showed how it could be done.

In fact, Spruce had not been the first to make such a proposal. That distinction belonged to Forbes Royle. However, it was certainly true that Spruce had been the first British botanist to actually arrive in the Andes and make his services available, and as Markham's own notebooks from the period make clear, he was fully aware of Spruce's proposal to the Foreign Office. Given that it was Markham who arranged for the Blue Books to be published in the first place, it is curious therefore that he did not see fit to include Spruce's letters in the parliamentary record. Perhaps there is an innocent explanation, perhaps not. The fact that Markham went to such lengths to petition for a pension for Spruce—and without Spruce's instigation—may be suggestive of a guilty conscience. On the other hand, he repeatedly drew attention to Spruce's pivotal role in introducing the red bark to India—hardly the action of someone who wanted to claim all the glory for himself.

SPRUCE WAS NOT the only one concerned about the historical record. From South America, Ledger had also been monitoring the progress of the cinchona plantations, and while he was proud that Mamani's seeds had turned out to be so productive, he was also beginning to resent the fact that he had not been afforded sufficient recognition for his role in the enterprise. Interestingly, Ledger also selected Howard as his confidant, but unlike Spruce, he was not nearly so shy about pressing his claims.

Ledger's extraordinary correspondence with Howard begins in 1874 and continues for almost nine years. His letters are by turns supplicating and boastful, and despite his attempts to disguise his motives, his resentment is frequently all too apparent. But when he isn't complaining to Howard about his "ill-treatment" by the British and Dutch, or soliciting his support for compensation, his letters are full of wonderfully vivid details of his adventures in the Andes and poignant anecdotes about Mamani.

All the strengths and weaknesses of Ledger's character are displayed in the first letter, written three years after Mamani's death. He had now abandoned Peru and moved to the province of Salta, in Argentina, where he had reverted to his former occupation as a general trader. As Ledger struggled to make a living dealing in alpaca ponchos and chinchilla hides, Howard—by now the owner of the most important quinine works in England and an esteemed member of the Linnaean and Pharmaceutical societies—became the sole conduit for his hopes and dreams.

"It is, as you may imagine, a source of great pleasure to me to see that the Cinchona seed forwarded by my aforementioned brother in 1865 has turned out so satisfactorily," he begins. "I am quite ignorant of botany. I have for many years taken and still take the liveliest interest in the successful cultivation of this most valuable tree in the dominions of Great Britain, rendering my native land independent for supplies from Bolivia or any other country . . . Had it not been for my kind brother, George Ledger, exerting himself, neglecting very likely his own business, carrying on a long correspondence; after all, this valuable seed in other hands would most likely have been lost."

The false modesty of these opening sentences is typical. In the next few paragraphs Ledger outlines the history of his expeditions in the bark forests of Peru and Bolivia, making it clear that though the insights into the location of the best *calisayas* were Mamani's, he was the first to propose putting the knowledge to good use. In 1845, he reminded Howard, he had written via his father to the British government, urging it to raise *calisayas* in Jamaica, India, and Ceylon so as to be independent of Bolivia in the production of quinine but had "never heard in what spirit my ideas were received." Next he had written to Markham, explaining where to go in Bolivia to collect the best *calisayas* and what precautions to take "so as not to excite suspicion as to his *real* object," but his letter had arrived too late to be of use. Ledger reminded Howard that Markham had never made it to Bolivia, yet in 1862 Ledger had put up the money for Mamani to travel to Apolo to collect the *calisaya* seeds. Having painted a moving portrait of Mamani's loyalty and devotion to the cause, Ledger concludes:

> After having witnessed the great precautions taken by different parties to "pack" the seed of the Cinchona for transmission to Europe, failing in almost every case to arrive in a "live" state, it is wonderful indeed how the "*seed*" sent by me in 1865 produces such splendid results . . . I cannot but feel that I have met with most inadequate acknowledgement after all my exertions, outlays and dedication of so many years. Surely, after the success attending the seed sent by me, the Govt of India and the Govt of the Netherlands should award me a sum of money commensurate with the importance and value of the service rendered.

There was some merit in Ledger's demands. The Dutch had paid him 100 guilders for the seeds and an additional 500 guilders when they germinated—a total of just 600 guilders (£120). Yet their plantations were now poised to reap huge profits. Similarly, although Ledger was not to know that the British would eventually squander his gift, Markham had to acknowledge that Ledger's expenditure in gathering the seeds had "far exceeded" the £50 he had received from Money for their sale. Further fueling Ledger's resentment was the knowledge that the Dutch had spent thousands on Hasskarl's expedition and showered him with honors, although his seeds had later turned out to be worthless.

At first, Howard was more than happy to help Ledger petition the Netherlands Indies government for compensation. In part this was because he recognized the force of Ledger's argument, but also because by now he could see that the *Cinchona ledgeriana* obtained by the British had been allowed to hybridize, and he wanted Ledger to obtain more seed for the Indian plantations. Unfortunately, Ledger's demands proved unrealistic—he wanted £2,000 to return to the *calisaya* forests—and Howard had to drop the proposal. Nevertheless, in the meantime, in an attempt to appease Ledger, Markham had also agreed to petition the British government for a £200 honorarium on his behalf.

Not everyone shared Howard's view of Ledger's contribution. Moens, the Dutch chemist, argued that Ledger's part in the success of the plantations was "very slight."

> The seeds have been gathered by an old servant of his, whilst he was on his travel back from Australia to Peru . . . If cinchona growers would make a statue for the clever cascarillero Manuel Incra Mamani, some reason could be found for that. But Ledger has indeed only a very small part in the history, and procured the seed without danger and without trouble, and without knowing indeed that he had got hold of such a precious kind of seeds.

Later Moens was even harsher, claiming that Ledger had built up his role in the enterprise only after he "knew how valuable a kind of cinchona he had been the means of procuring."

Moens went too far. Ledger may not have risked his own life to collect the seeds, but he certainly risked his money, and there is little doubt that he knew their potential. Moreover, in 1879 he discovered that the Dutch had offered to pay one of his former clerks £400 for twenty-five pounds of seeds of equal quality to those his brother had sold the consul in London in 1865. It was with some justification, then, that Ledger concluded that he had not been treated "in good faith or honesty."

As the exchanges became more heated, Howard desperately tried to find a middle way. He argued judiciously that while botanists owed their knowledge of the best kind of *calisaya* bark to "the experience and practical sagacity of an Indian," it was Ledger who deserved "the whole credit of the enterprise of obtaining the precious seed to which the hope of future success attaches in Java and perhaps in other parts of the East Indies."

However, it was Moens who won the initial skirmish. Rejecting Ledger's petition for further compensation, the Dutch in 1880 agreed to award him a premium of 1,200 guilders (£242). The British were even more peremptory, rejecting Markham's appeal for a £200 honorarium for Ledger out of hand.

Ledger tried to disguise his disappointment. He waived his claim for compensation against the British and wrote to Howard, "I am delighted to hear that fine Ledgeriana are at last! thriving in India and Ceylon. Hurrah! Three times three!" However, in almost the same breath Ledger made it clear that it was Mamani who had got "the true calisaya," and that he was determined to continue petitioning the Dutch.

Eventually, his nagging paid off, although not quite in the way he expected. Howard had already begun referring to the seeds as *Calisaya ledgeriana.* Now, in 1881, Moens, perhaps in response to Howard's representations and wishing to find some way of placating Ledger, formally christened the species *Cinchona ledgeriana.*

In botanical terms, at least, it was the ultimate accolade. At long last, Ledger had received the public recognition he craved, and in writing to Howard in 1881 he seems cheerfully resigned to his place in history. "I am indeed glad, very glad, to see Cinchonas being so successfully cultivated in so many parts of the world and to see my humble name handed down to posterity as one of the helpers in the great work."

If Howard thought he had heard the last of Ledger's demands, however, he was very much mistaken. For of all Ledger's talents, his greatest was his capacity to bear a grudge. Had his other schemes for making money ever borne fruit, he might have forgotten the matter. But Ledger's tragedy was that he never found a product as valuable as Mamani's seeds. And when, like Spruce, he lost everything in a banking disaster, he had nothing else to exploit.

THE BRITISH HAD NEVER ENTIRELY given up hope of matching the output of the Dutch plantations. Although the seeds that George Ledger had sold to Money had been allowed to hybridize in India, he had also given some to Howard. These Howard had carefully nurtured, first at his greenhouse in Tottenham and later in the nurseries at Kew Gardens. Beginning in 1875, Kew began sending cuttings raised from these original seeds to Ceylon and Jamaica. But the soil and altitude there must have been different from Java's, or else the Dutch management of their plantations was more scientific. Either way, the Javanese plantations continued to produce twice as much quinine per acre as the British, and gradually the India Office realized there was no point trying to compete. In the future, the Nilgiri and Sikkim plantations would cultivate bark solely for the production of quinetum, thus meeting the domestic needs of British subjects in India. Meanwhile, British demand for quinine outside India would be met either by leasing plantations from the Dutch in Java or by purchasing quinine on the open market.

From both a political and commercial point of view the compromise made perfect sense. Malaria was such a widespread problem, and the challenge of cultivating cinchona so difficult, that from the beginning both countries had been happy to exchange seeds and scientific know-how in search of a lucrative strain. Now that the Dutch had succeeded and the price of quinine was beginning to fall, there seemed little to be gained from holding out for more. Quinetum was now available at any post office in Bengal for half a farthing a dose. The truth was that even at this highly subsidized price it was still beyond the means of most Indians. Contract coolies, and their wives and families, in high-risk areas could afford it. But for the self-employed or indigent, half a farthing was too high a proportion of their income. The conclusion was clear. There was a limit to philanthropy, and in India the British had reached it. If any profits were to be made from the bark in the future, they would come from selling quinine to those who could afford it—in other words, "rich" Europeans and Americans.

If Ledger had not been living in isolation in Argentina, he might have seen the logic of these arguments and saved himself a lot of grief. Instead, displaying the same misjudgment that had plagued the alpaca venture, he invested the money the Dutch had awarded him in buying the latest quinine preparations from Howard. It was, of course, a disaster: in 1880 the price of quinine began to slide precipitously, and by 1881 the margins Ledger had been hoping to trade on had evaporated.

There was now little to keep Ledger and his wife in South America. In letters from friends and family in Peru and Bolivia, he learned that animosity to him in those countries was growing. Because of the success of the Dutch plantations, made possible by the seeds he had "stolen" from Bolivia, Ledger was now being accused of having "destroyed a trade employing 20,000 persons" and "depriving many of fortune."

From Ledger's point of view, he had not gained anything. Although he had cleared his debts, his only assets were the few alpacas he had brought with him from Bolivia. He and Charlotte now decided that there was little point in remaining in Argentina. They would return to Australia, where they still had family and the prospect of making a living. When Ledger's father died in 1883, and it became clear that George's health was not good either, he made up his mind. He would catch a steamer to London for the funeral reception and then, after a last visit to the family home, immigrate to Australia.

Ledger must have had mixed emotions as he embarked at Buenos Aires. He was now sixty-five and had not set foot in Britain for forty-seven years. In 1836, when he set sail for Peru on the *Swallow*, he had been full of youthful enthusiasm. Now, as he crossed the equator once again, he had plenty of time to reflect on what could have been, if only he had possessed the "right" connections. More than anyone else, Markham had been the key. If only he had known about his mission sooner, or his letter had not missed him at Tacna, the British might now be enjoying the fruits of his quinine-bearing seeds and he would be a rich man.

It would be interesting to know if Ledger expressed any of these regrets when he reached London and was finally introduced to Markham at Howard's office in Tottenham. Howard had also invited Moens, but he declined. He did not like Ledger—that much is clear.

There is no record of the meeting, but in all probability Ledger kept his thoughts to himself. Markham had been unstinting in his support, and there would have been little point in antagonizing him. In his book on cinchona cultivation, Markham had written that although Ledger had not "served under" him, as soon as he heard about the mission he had "set to work, actively and zealously, to co-operate in the good cause."

Markham could afford to be generous. Unlike Ledger, cinchona had made his name. In 1868, on the strength of his success in Peru and India, Markham had been appointed official geographer to Robert

Napier's expedition to Abyssinia. Other honors soon followed: he was made a Fellow of the Royal Society and Honorary Secretary of the Royal Geographical Society. In 1872, with the help of Kew, he had dispatched collectors to the Amazon to seize another valuable botanical product of South America—*Hevea* rubber, an assignment that Henry Wickham completed four years later, when he managed to smuggle seventy thousand *Hevea* seeds onto an English steamer at Belem. Next, Markham began promoting polar exploration, joining an expedition to Greenland in 1875. He had dedicated himself since then to historical writing and was now encouraging expeditions to the South Pole.

Ledger's subsequent letters to Howard mainly went over old ground, discussing how the British had made a hash of cinchona cultivation and had ignored all competent advice. Previously, Ledger had told Howard how much he appreciated Markham's support for those who had "served under him," and in all probability he repeated similar sentiments in his presence. Secretly, however, Ledger could not forget that he had been to Bolivia, whereas Markham had not.

But there was little point dwelling on that now. The meeting had convinced Ledger there was nothing further for him in Britain and that he should leave for Australia immediately. Two days before embarking, he wrote to Howard one more time to thank him for his "kindness and support." The letter concludes: "Something may 'turn up' and indeed it would be most welcome . . . As before repeatedly stated I'm no 'Botanist'—the little I know has been acquired practically and from what I heard from my faithful old servant and friend Manuel and other most intelligent Bolivian Indians."

It was the end of their correspondence. Shortly after Ledger's departure, Howard died. From now on, only Markham could help him.

10

The Great White Hope

Without exaggeration, and without excessive optimism, we must continue to get quinine used as much as possible.

—ANGELO CELLI

ONE OF THE MOST REMARKABLE ASPECTS of the story of cinchona is that for 250 years everyone knew that the bark cured malaria but no one had the faintest idea why. Despite spending fifteen years in the Amazon and Andes, for instance, Spruce never once seems to have questioned the medical orthodoxy that malaria was the result of "poisonous exhalations" from the land. Nor does he seem to have asked himself why cinchona, rather than any other plant, should block the fever cycle. Like Wallace, Darwin, and other eminent nineteenth-century naturalists, Spruce simply had no concept of the germ theory of disease. Nor did he have much interest in plant pharmacology or chemotherapy. As far as Spruce was concerned, malaria was simply an intractable fact of life—the cost, if you like, of working in the tropics—and quinine another of nature's imponderables.

While Spruce, Ledger, and Markham were hunting for cinchona, however, another group of malaria pioneers was following a parallel trail. Like the naturalists, they would need determination, tenacity, and a fair degree of luck. The difference is that they could do it all from the comfort of a laboratory bench.

The life cycle of the malaria parasite is complicated and involved, but for present purposes the only bit that need concern us is the merozoite, or so-called ring stage. This is when the young parasites emerge from the liver and suddenly begin feeding on red blood cells from the inside. Thanks to powerful electron microscopes, we now know that the merozoite burrows through the wall of the corpuscle and begins avidly devouring the cell's protein—the hemoglobin. In the process, however, it also releases the iron-bearing heme and the hemazoin it produces, which is toxic. To stop itself from being poisoned, the merozoite locks the hemazoin away as malaria pigment. Quinine interrupts this process by binding to the heme in a manner that is still not fully understood and preventing the production of the malaria pigment. In other words, it causes the young parasites to choke on the products of their own digestion.

If you know what you are doing, this process is easily seen through a microscope. The malaria pigment appears as a black dot or dark granule within the circular bodies of the merozoites. If quinine is not introduced, and the merozoites are left to follow their natural course of development, some of them grow so big as to fill, or almost fill, the entire red blood cell. Next, they divide, emerge from the blood cells, and invade fresh red corpuscles, sparking the fever attack. Eventually, after several such cycles, some merozoites cease dividing and develop instead into sexual-stage gametocytes. Viewed through a microscope, the gametocytes appear as crescents, in the case of falciparum malaria, or spheres, in the case of other types of malaria.

At first these gametocytes just linger in the blood, waiting for another female *Anopheles* to bite. Once reingested by the mosquito, the males undergo a further spectacular transformation—sprouting flagella, or filaments. These are the sperm, the male gametes that will fertilize the female gametes inside the mosquito's stomach, nearly completing the life cycle of the parasite.

Microscopes have been around since the end of the seventeenth century—in other words, at least as long as physicians have known about cinchona. However, it was only when Louis Pasteur's and Robert Koch's groundbreaking work on bacterial infections renewed interest in microscopy that scientists began taking a closer look at the blood of malaria patients. Once they did, it was simply a matter of time before someone stumbled on the truth. The dark, iron-bearing malaria pigments are like droppings along a trail. If you follow their path, quite soon you will discover the characteristic forms of the parasite. And if you happen to be looking at fresh blood from someone with a malaria infection, you might even see them moving.

The first person to notice the pigments was a German psychiatrist, Heinrich Meckel. In 1847 he was examining the blood of a patient with an enlarged spleen when he recorded a series of "black irregular rounded granules united . . . to globular, egg-shaped or fusiform bodies." Meckel did not mention whether the patient had intermittent fever, but soon other researchers were taking note of the pigmentation, and gradually doctors began to associate "melanism," as the phenomenon became known, with malaria.

In 1858 the German pathologist Friedrich Theodor von Frerichs demonstrated that the black particles resulted from the destruction of human red blood cells. Then, in the 1870s, Achilles Kelsch, a French army pathologist at the Val-de-Grâce military hospital in Paris, noted that the pigments were formed at precisely the time the fever came on.

However, the real breakthrough came in 1880, when another French army doctor, Charles-Louis-Alfonse Laveran, in Algiers, decided to take a closer look at the pigments. Like Kelsch, Laveran had only a very basic microscope, but unlike Kelsch, Laveran got lucky. On November 5, 1880, a twenty-four-year-old soldier of the 8th Squadron of Artillery presented himself to Laveran, complaining that he was still suffering fever and chills despite having been treated with quinine three weeks earlier (he had probably been given an inadequate dose). Laveran had already found a large number of crescents—the gametocytes of falciparum—in the soldier's blood. Now, before dosing him with quinine again, he took another blood sample and placed it on a slide. This time Laveran not only saw the crescents but, astoundingly, a number of spherelike bodies with moving filaments. It was the first time anyone had observed the malaria parasite—in this case, in its sexual stage. "I was astonished to observe that at the periphery of this body was a series of fine, transparent filaments that moved very actively and beyond question were alive," Laveran later recalled.[1]

Incredibly, Laveran's discovery was ignored for four years. Despite his having clearly seen the parasite, his observations were dismissed mainly because most leading scientists were convinced that malaria must be caused by a bacillus like the ones discovered by Pasteur and Koch. Indeed, it was not until the invention of the oil-immersion lens microscope in 1884 and improved aniline stains brought the parasite into sharper focus that the Italians—the arbiters of malaria research—began to take Laveran's findings seriously. In autumn 1884 and throughout 1885, the Italian researchers Ettore Marchiafarva and Angelo Celli patiently examined hundreds of blood slides, assembling what was, in effect, a series of freeze-frames of the parasite's development. At the same time, another Italian, Camillo Golgi, observed that the parasites seemed to be synchronized—they divided almost simultaneously and at regular intervals—and that their division coincided with the onset of fever. These observations would eventually lead, in 1892, to the conclusion that the crescents and rings seen by Laveran twelve years earlier were different forms assumed by a single species of parasite, and, in 1897, to the realization that the flagellating gametocyte was the same parasite in its sexual form.

As far as quinine's reputation was concerned, however, there was no need to await these further revelations. Although the scientific community was slow to accept Laveran's conclusion that malaria was a parasitic illness, his observations provided clear and dramatic proof that quinine stopped the pigmentation process in its tracks. Laveran was still a long way from explaining how quinine achieved this remarkable feat, but you only had to look at the blood of a malaria patient who had been treated with quinine to see that the black pigments disappeared. In the future, quinine could no longer be viewed merely as a therapeutic, a drug for easing fever. To use the sort of military terminology that Laveran—who, after all, was an army doctor—might have appreciated, it was now a weapon, a magic bullet against the disease itself, and an essential part of the arsenal deployed by Europeans in the tropics.

But while the microscope broadened quinine's horizons, it also, for the first time, marked the drug's limitations. Following Laveran, it soon became clear that quinine had no discernible effect on the crescents and spheres.[2] The gametocytes persisted no matter how much you dosed them. In other words, quinine killed the parasites in the blood cells, but it had no effect on their sexual form and thus could not be considered a complete sterilizing treatment that could prevent transmission. "Even, when giving quinine every day by subcutaneous injection, and in large doses for a whole month," wrote Celli in 1901, "I have not often succeeded in destroying the crescent forms . . . upon these forms of the parasite, which are the most dangerous from the epidemic point of view, quinine has no influence . . . therefore a complete rational prophylaxis by means of disinfection . . . is not easy, nor is it always possible."

However, for quinine's limitations to become apparent to Celli, another army doctor, the British surgeon-captain Ronald Ross, first had to made the crucial link between the parasite and the mosquito. This Ross did in 1897, dissecting an *Anopheles* in Secunderabad and discovering the plasmodium (in its oocyst form) hidden in its stomach wall. But not only did Spruce, Ledger, and the others not know about the plasmodium; they did not understand the mosquito's role with respect to it.

To be fair, no one did, though there had been suspicions. The earliest reference to the mosquito as a vector comes in an Indian text known as the Susruta. Written by a Vedic scribe in the sixth century B.C., it refers to five kinds of disease-carrying *masakah* (mosquitoes) whose bite is accompanied by, among other symptoms, "fever, pain of limbs . . . shivering . . . burning sensation, intense cold." The Babylonians portrayed their pestilence-causing god Nergal as a two-winged fly, while the Canaanites called him Beelzebub, "prince of flies."

Similarly, although the association between malaria and "bad air" is an Italian idea, some classical Roman writers dropped tantalizing hints that there might be more to it than that. Varro in his *De Re Rustica* posited that invisible microorganisms breeding in marshes and stagnant waters might be responsible for the disease and recommended that houses be built in high and well-ventilated places so that the small "*bestiolae*" would be blown away. And writing at the end of the first century A.D., Columella pointed out that marshes bred "animals armed with mischievous stings, which fly upon us in exceeding thick swarms; and also sends forth, from the mud and fermented dirt, envenomed pests of water snakes . . . whereby hidden diseases are often contracted."

Unfortunately, so great was the Roman obsession with the foul air emanating from swamps that these tantalizing glimpses of the truth were ignored, not least by Hippocrates, who was convinced that the cause was stagnant water. By the seventeenth century—when malaria became widespread in Europe—the association between malaria and the insalubrity of marshes was well established. The French term for malaria, *le paludisme*, comes, as does the Spanish *el paludismo*, from the Latin word *palus*, meaning "swamp." Similarly, although the English term *ague* refers to the characteristic swelling of the spleen in malaria patients (the so-called ague cake), in *The Tempest*, Shakespeare has Caliban call forth "All the infections that the sun sucks up/From bogs, fens, flats on Prosper fall and make him/By inch-meal a disease." While Walpole referred to a "horrid thing called mal'aria" on his return from Rome in 1740, it was only with the publication of John MacCulloch's *Malaria: An Essay on the Production and Propagation of this Poison* in 1827 that *malaria* became the preferred term in English medical circles. That cemented the association between the disease and the chemical composition of the air, but doctors were no nearer to understanding the real pathology of malaria and thus had no accurate way of diagnosing it, other than by their observations of the periodicy of the fever attacks and patients' response to treatment with quinine.

Because of this, there are no accurate figures of the disease's incidence until the early 1900s, when the advent of blood tests for the malaria parasite made it possible to confirm clinical diagnoses. However, we know from anecdotal reports of intermittent fevers and agues that malaria was widespread in Europe and North America. For instance, between 1783 and 1794 "ague" accounted for half of all admissions to St. Thomas's Hospital, in Lambeth—then one of the most malarial areas of London. And throughout the nineteenth century Holland's North Sea coast suffered repeated epidemics of "malarial disease," one of the worst occurring in Groningen in the summer of 1826, when the sea burst through the coastal defenses, flooding the countryside and resulting in eight thousand fever cases and three thousand deaths out of a population of thirty thousand. Similar epidemics occurred along Germany's northern seaboard and in the Lorraine, Gironde, and Lande regions of France in 1846 and again in 1849. Indeed, between 1865 and 1867, deaths from intermittent fevers throughout France were running at six in one thousand. Italy was more affected. In 1887, the first year for which figures are available, twenty-one thousand deaths were attributed to malaria, and by the turn of the century it is estimated that one in fifteen Italians, or some two million people a year, were contracting the disease.

Nor was the situation much better in the United States. New York and Philadelphia had long been malarial, but by the early 1800s, settlers migrating west had introduced the disease to Illinois. And during the American Civil War, malaria accounted for 1.3 million casualties and ten thousand deaths on the Union side alone.

The first person in the modern era to begin to suspect the mosquito's role in transmitting malaria was another French doctor, who worked, like Laveran, in the tropics. More intriguing, this doctor was based in Caracas, Venezuela, less than five hundred miles from the Orinoco, and published his theory just months before Spruce succumbed to fever on the river.

Born in Guadeloupe in 1803, Louis Daniel Beauperthuy had a strong biological bent. Like Laveran, he enjoyed nothing better than looking at blood through a microscope. After studying medicine in Paris, he had returned to his native Caribbean to be closer to the seat of such diseases as leprosy, yellow fever, and malaria.

Whenever an epidemic broke out, Beauperthuy would study it on the spot, and in 1854 his work as a health officer for the Venezuelan government brought him to Cumaná, then in the grip of a yellow fever outbreak. Beauperthuy was familiar with the miasma theory of malaria but could not help noticing that though Cumaná was humid, there were no swamps. There were, however, plenty of mosquitoes. Noting that the Roman historian Herodotus had referred to mosquitoes as winged snakes, Beauperthuy decided to take the idea a stage further. In May 1854, as Spruce was making his way downstream to Maipures, Beauperthuy wrote in the *Gaceta Official de Cumaná*:

> The mosquito plunges its proboscis into the skin . . . and introduces a poison which has properties akin to snake venom . . .

> Marshes do not communicate to the atmosphere anything more than humidity, and the small amount of hydrogen they give off does not cause in man the slightest indisposition in equatorial and inter-tropical regions renowned for their unhealthiness. Nor is it the putrescence of the water that makes it unhealthy, but the presence of mosquitoes.

Although Beauperthuy rather ruined his theory by conjecturing that the source of the "poison" introduced by the mosquito was decomposing animal matter, it was a stunning leap, and soon others were edging toward similar theories. By the middle of the nineteenth century in the United States, malaria was rampant along all the watercourses from the Mississippi to the Potomac. However, a number of American doctors had begun to notice that malaria was also prevalent in the hill country, where marshes and stagnant water were absent. In 1848, Josiah Clark Nott, a doctor from Alabama, had tentatively raised the possibility that the culprit might be the mosquito because of its ability to "take flight." However, it was not until 1882 that another doctor was prepared to be more outspoken.

Albert Freeman Africanus King was an Englishman who went to America as a boy, became an obstetrician, and practiced for nearly fifty years in the nation's capital. Washington, D.C., and the surrounding peninsula had been so malarious during the Civil War that during the march on Yorktown General McClellan's Army of the Potomac had been stopped in its tracks by the disease. The effects became particularly noticeable when the Union troops camped on the banks of the Chickahominy River and were beset by "Chickahominy fever." Twenty years later, it was no better. The United States Sanitary Commission had already requested a report on how to tackle Washington's "miasmatic fevers." Now, in a paper before the Philosophical Society in 1882, King listed nineteen reasons he believed mosquitoes were the vectors of malaria, and challenged the city's fathers to erect a massive wire screen around the capital as high as the Washington Monument to test his theory.

Unfortunately, King's novel suggestion was dismissed. His distinguished medical colleagues still considered the mosquito theory too fantastic, and King is best remembered today for attending Lincoln on the night he was assassinated at Ford's Theatre. The one thing that could have given King's musings credibility was Laveran's theory about the parasite. But though Laveran had published his findings three years earlier, they had met with such indifference that few doctors in America, including King, it seems, were aware of them.

It was the British who were destined to provide the crucial link between the mosquito and the parasite, and they did it not through an intuitive leap but by steady, laborious experimentation. The first step was taken by Patrick Manson, a medical officer in the Chinese Imperial Customs Service. In 1876, while dissecting mosquitoes in Amoy that had fed on human volunteers infected with elephantiasis, a parasitical infection of the filarial worm, Manson discovered filariae in the mosquitoes' body cavities. However, although he studied filariasis for several years, he never followed the parasites to full maturity and thus never saw them proceed to the mosquito's head, enter her proboscis, and at the moment of biting, slide down the puncture into a new human host.

Manson's failure to take his studies to their logical conclusion led him into a cul-de-sac that could easily have prevented Ross, his disciple, from ever discovering the true transmission process. On his return to England, Manson had reasoned that mosquitoes were simply nursemaids to the filariae, and that the parasite passed to humans only when they drank from water infected with filariae-containing mosquito larvae. Nevertheless, Manson inspired Ross sufficiently to persuade him to invest in a microscope and begin examining mosquitoes for signs of the malaria parasite.

A reluctant medic whose first love was poetry, Ross seems to have had little understanding of parasitology or mosquito taxonomy. However, he was a dogged researcher and, stationed in a remote, malarious Indian outpost, found himself ideally placed and with plenty of time to pursue the insect vector. Shortly before he made his breakthrough at Secunderabad, Ross had passed by the cinchona plantations at Ootacamund—or "Ooty," as he liked to refer to the hill station. At the time, Ross had been on his way to the highly malarious Sigur Ghat to trawl the pools and lagoons for mosquito larvae, and in preparation had been dosing himself with five grains (325 mg) of quinine daily. But quinine, as we know now, is a poor prophylactic, and when he reached the ghat, Ross suffered a "severe go of fever." To cure himself, he upped the dose, but he later recalled that the experience left him "weak and depressed for a long time."

Indeed, it was only while straining his eyes one steamy summer evening back in Secunderabad that the young military physician's spirits finally lifted. Ross later referred to that day—August 20, 1897—as Mosquito Day. That was when he saw for the first time a spherical body (oocyst) on the stomach wall of an *Anopheles* that had previously fed on a malarious patient. He recognized that it contained pigment identical to that seen in the malaria parasite. The following July, in Calcutta, while examining bird malaria transmitted by the *Culex* mosquito, Ross saw the oocysts burst, invade the insect's thoracic cavity as "threads," and migrate to the mosquito's salivary gland ready to be inoculated. As far as Ross was concerned, the *Culex* was all the proof he needed. "Hence I think I may now say QED and congratulate you on the mosquito theory indeed," he wrote to Manson.

The congratulations were a little premature. Ross never seemed to have fully grasped the fact that it was only a certain type of mosquito that transmitted malaria in humans, and he never went on to examine whether anophelines—or "dapple-winged" mosquitoes, as he called them—also contained oocysts and threads. That proof was left to Giovanni Grassi, professor of anatomy at Rome University. But unfortunately for the Italians, Grassi's paper showing that human malarias were transmitted only by species of *Anopheles* was not published until four months after Ross had announced his discovery to the British Medical Association. So it was the Englishman, not the Italian, who in 1902 was awarded the Nobel Prize.

WHAT SPRUCE WOULD HAVE MADE of Ross's breakthrough had he lived to see it is anyone's guess. Perhaps he would have found irony in the fact that while he had been struggling to sit up straight at his own microscope to put his collection of mosses and liverworts in order, Ross had been using the same instrument to such dramatic effect.

Spruce's doctor's belated diagnosis of a "stricture of the rectum" had done little to ease his suffering. In 1869 he wrote to a friend from Welburn that "sitting up to the microscope has brought on bleeding of the intestines to such an extent that I fear I must renounce the task altogether." For the next eighteen months he was in too much pain to use his microscope at all, and it was only in the spring of 1871 that he "plucked up the courage" to disinter it and resume work on his hepaticae.

In 1873, Spruce moved to a small cottage in Coneysthorpe, just outside the gates of Castle Howard. The sitting room was just twelve feet square and the bedroom even smaller. Although he had a housekeeper and young girl to nurse him, and the Howard family looked in on him, his options were few. Too infirm and poor to contemplate further travel, all his efforts were now devoted to organizing his massive collection. It was meticulous, painstaking work, but as Wallace would later recall, Spruce was "a man who, however depressing were his conditions or surroundings, made the best of his life," and by 1885 his six-hundred-page work on Amazonian hepaticae was complete.

Of all the areas Spruce had visited, it was the Río Negro that had enthralled him most and where he had discovered a great number of the species and genera new to science. But he had also been captivated by the Andean cloud forests, not least because the same damp, clinging mists that were conducive to cinchona also favored the growth of mosses, lichens, and fungi. Indeed, plant pharmacologists believe today that the alkaloids in cinchona bark probably evolved in response to this mossy environment, their bitterness helping to discourage fungi and other plant predators. But if such a thought ever occurred to Spruce, he never recorded it. Although he liked to think of himself as a Darwinian, his response to plants was first and foremost aesthetic. "I like to look on plants as sentient beings," he once wrote in response to some deprecatory remarks about his beloved hepaticae.

> It is true that the Hepaticae have hardly as yet yielded any substance to man capable of stupefying him, or of forcing his stomach to empty its content, nor are they good for food; but if man cannot torture them to his uses or abuses, they are infinitely useful where

> God has placed them . . . they are, at the least, useful to, and beautiful in, themselves—surely the primary motive for every individual existence.

Spruce too had been "infinitely useful" in Ecuador, the place where fate had placed him. After all, it was thanks to Spruce that Ross, and thousands of other malaria sufferers in India, survived to lead productive lives. Moreover, the trail Spruce had followed had been fraught with real danger, whereas the only pitfalls Ross had risked were professional ones. But there was no Nobel Prize for Spruce, none of the acclaim that accompanied great scientific achievement. Instead, in 1864 he received an honorary doctorate in philosophy from the Imperial German Academy in Berlin, and two years later, an honorary fellowship from the Royal Geographical Society. Then, in 1893, he was elected an associate of the Linnaean Society.

It was too little, too late. Spruce was now seventy-six and so incapacitated that he could barely walk, let alone write. Four years previously he had suffered another paralytic attack, as a result of which he had been unable to leave his cottage for two months. Now, shortly after Christmas 1893, he contracted influenza, and after a severe attack, he died on December 28.

Spruce was buried at Terrington, the parish in which he had been born, beside his father and mother. The Countess of Carlisle sent a wreath from Castle Howard as a token of remembrance, but the only people present at the graveside were Matthew Slater, the executor of Spruce's estate, and a few close friends from Coneysthorpe.

After Spruce completed his book on Amazonian hepaticae, his last years had been spent working on British and European hepaticae. He had no strength left to write a popular account of his travels, a task that now fell to Wallace. Fortunately, Spruce was a conscientious and descriptive diarist, as well as a prodigious letter writer, and under Wallace's skillful editing—informed by his own experiences in the Amazon—an enduring testimonial emerged. Spruce may not have understood malaria or how quinine worked, but he was one of the last great naturalist-explorers. And from the botanical notes and journals he kept in South America emerges a vivid picture not only of the plant life and geography of the Amazon and Andes, but of the customs and habits of the indigenous peoples among whom he spent the most arduous and, arguably, the best years of his life.

Given his reputation as a collector, the only surprise is that Spruce never discovered a new species of cinchona. Instead, he is commemorated in the names of two genera of mosses and liverworts—respectively, *Sprucea* and *Sprucella*—and two species of British liverwort, *Marsupella sprucei* and *Orthotrichum sprucei.*

IN CONTRAST TO ITS ROLE in Spruce's life, disease does not appear to have figured much in Ledger's life or his imagination. During his travels over the cordillera into Bolivia and Chile, he suffered altitude sickness and no doubt endured his fair share of dysentery. But in all his time in South America he never once refers to an intermittent fever or having to take quinine. And even later in his life, after Ross's discoveries had been announced to the world, there is no record in Ledger's letters of so much as a passing interest in the pathology of malaria—the disease that his seeds had done so much to alleviate.

In 1891, however, like Spruce two years later, Ledger succumbed to influenza. The epidemic fell upon him and his wife in Goulburn, in New South Wales. He and Charlotte had bought a small farm there with money from his father's inheritance and the sale of the remaining alpacas in Argentina. The plan was to retire on their modest savings—some £1,500 in all, which Ledger had deposited in a local branch of the Australian Banking Company. But the flu was a severe one and Charlotte died. An obituary notice in the local paper mentioned that Ledger, too, was "very low and not expected to live throughout the day."

That notice would later lead to rumors in Britain that Ledger was also dead. In fact, he survived, but in some ways it may have been better had he not. The very day that Charlotte's obituary notice appeared, another article in the same paper announced that the Australian Banking Company had "suspended payments." The fall in the price of wool, coupled with the embezzlement of funds by a senior director, had eaten into all the bank's deposits. A year later, the bankruptcy was confirmed, and Ledger—like Spruce thirty years earlier—found himself bereft of savings.

Now, aged seventy-three, Ledger was broke and all alone. His only hope was Markham. Following the success of the Indian plantations and the Napier mission to Abyssinia in 1868, Markham had been showered with honors, and in 1893 he was elected president of the Royal Geographical Society. If anyone could secure him a pension from the British and Dutch, reasoned Ledger, it was Markham.

Once again Ledger put pen to paper, and once again Markham renewed the appeals, writing a long letter to the *Chemist and Druggist* outlining Ledger's plight. To the eminent quinologists and chemists who subscribed to the specialist weekly, the news that Ledger was alive and broke in Australia came as a surprise. "A lost cinchona-pioneer found," announced the headline. A few weeks later, the *Chemist and Druggist* went a stage further, writing a long editorial in support of Ledger's claim and calling on planters, importers, and quinine manufacturers everywhere to contribute to a fund on his behalf.

Ledger had been "very shabbily rewarded," opined the editorial. "We doubt whether, as a matter of bare commercial principle, it would be possible for Mr Ledger to show any cause why either the Dutch or Indian Governments, to whom, for a hundred pounds or so, he supplied seed that has produced cinchona bark worth many thousands of times the sum paid for it, should do anything for him in his old age. Fortunately, however, for those in distress, sentiment counts for something in the governments of the world . . ."

It was the same argument Markham had put to the secretary of state for India in 1880, and the unsentimental British were no more inclined to accept it now than they had been in the past. At first the Dutch were not moved either. At this point Ledger was asking only to be reimbursed for the monies he had spent procuring the *ledgeriana* seeds. But the Dutch rejected the request out of hand, saying he had already received "repeated liberal payments"—a reference to the total of nearly £400 he had been paid since 1865. The phrase infuriated Ledger.

"Had I been born in Holland I should have been covered with medals and be a wealthy person now," he fumed, adding that by refusing to compensate him, the Dutch were condemning him "to end my days in an asylum for the destitute."

It was hyperbole, of course. By 1895, Ledger had mortgaged his cottage and was living on the interest from the money raised. But his income was only £3 a month and the *Handelsblad* of Amsterdam, one of the leading Dutch daily papers, was sufficiently appalled to take up his cause. Calling on the Dutch government to grant him a pension of £100 a year, the paper began a sustained campaign. Markham quickly pitched in, adding his weight as president of the RGS by writing directly to the Dutch minister for the colonies.

At first the Netherlands Indies government resisted, but gradually, under the weight of public pressure, it was forced to give in, and in 1896 the Dutch parliament voted Ledger a £100 pension "for distinguished services rendered by him to the cinchona industry." On May 4, 1897—Ledger's seventy-ninth birthday—the Dutch minister for the colonies formally notified him of the annuity. After more than twenty years of canvassing and complaining, Ledger's fight was finally over.

In the last photograph of him, taken shortly before his seventy-ninth birthday, he is shown seated in a Victorian parlor, with one foot propped on a cushion and his top hat resting upside down on his writing desk. His left hand is still gloved, as though he has just come in from a brisk walk, while his lips are clamped tightly shut, giving him a stern and forbidding look. The portrait was probably meant to convey Victorian respectability, but his eyes seem to burn with bitterness, and it is hard not to see in his piercing gaze the unrequited resentment for all his misfortunes.

Very little is known of Ledger's old age. He lived eight more years, finally dying in May 1905 at the age of eighty-seven of bronchitis. He was buried beside his Australian sister-in-law, Emma Garratt, in Rookwood Cemetery, a sprawling Methodist burial ground outside Sydney. At the time of his death, he was so poor that his name did not appear on the tombstone.

BY THE TURN of the century malaria had begun to recede from northern Europe, but in southern and eastern Europe and the tropics it remained a major cause of morbidity and mortality. In 1902, for instance, Italy reported 178,000 infections and nearly 10,000 deaths; in 1905 the number of infections was even higher: 323,000. Quinine represented the Great White Hope. With abundant and cheap supplies of the drug flowing out of Java, it seemed only a matter of time before malaria was eradicated from Italy and every other corner of the globe, making malarial zones safe for whites who had not acquired Africans' immunity to the disease. Quinine's biggest supporter was Robert Koch, the German bacteriologist who had been one of the most skeptical critics of Laveran's claims to have seen the parasite. In New Guinea, Koch had seen for himself how efficiently quinine cleared the parasite from patients' blood, and he now became the leading advocate of quinine prophylaxis, a method he believed "could make every malarial place free or almost free of malaria" within nine months.

The Italians concurred. Some, like Celli, went even further. "Without exaggeration, and without excessive optimism," he wrote in 1901, "we must continue to get quinine used as much as possible; we must administer it promptly, assiduously, and in greater quantities than Koch prescribed."

But quinine alone would never be the complete answer to malaria, for several reasons. First, as Celli had already observed, it did not act on the gametocytes; nor did it affect the growing liver stages of the parasite.[3] Second, if used for a prolonged period, it had unpleasant side effects, including headaches, nausea, ringing in the ears, and, in cases of extreme falciparum infection, blackwater fever—an autoimmune response in which hemoglobin suddenly breaks down, staining the urine a dark red or black. Third, and perhaps most important, quinine's "affordability" could not last: sooner or later the economics of bark production in Java would force up the price.

However, by now the identification of the mosquito as the culprit had raised the possibility of complementary strategies—ones that in tandem with quinine could deliver mankind to the promised land of a malaria-free world. In 1900, Manson oversaw a series of experiments on the Roman Campagna in which he demonstrated that it was possible to block the transmission of the parasite from mosquito to man through the simple expedient of screens. In India, meanwhile, Ross had done complex calculations to show that if you filled in enough ditches and greased enough ponds and puddles with oil, you could stop *Anopheles* from breeding. Then there was the Italian idea of extensive public sanitation works that drew inspiration from Rome's ancient *cloaca maxima.*

In 1901, at a conference on malaria in Eastbourne, England, Koch—just returned from New Guinea—outlined a four-point plan for eradicating malaria. Pride of place still went to quinine, especially in the tropics, but the use of nets and screens, the destruction of mosquito breeding grounds, and the removal of malarious patients were now almost as important. A year later, the Italian government decided to follow "Koch's Way," establishing a state quinine service to subsidize production and distribution of the drug. In the first year of operation alone, the government sold more than two tons, chiefly in sugar-coated tablets, and by 1907, sales exceeded 23 tons. At the same time, it began a vigorous campaign of bonification, draining swamps and marshes and filling in unsanitary ditches. The result: deaths from malaria in Italy fell from a high of fifteen thousand in 1900 to just over two thousand by 1914.

In India, "Ross's Way" took precedence. At Mian Mir, a notoriously malarial plateau in Lahore, the British army cleared and oiled irrigation ditches, exported infectious patients, and administered quinine in both preventive and curative doses. Unfortunately, the results were not encouraging. By 1909, two thirds of local children were still being infected—as many as when the experiment had begun—and General Kitchener decided to stop the trials. Meanwhile, in India as a whole, malaria cases were still running at a hundred million a year. The difference was that, thanks to quinine and quinetum, by 1910 deaths were down to one million—half the mortality level of fifty years previously when Markham began cinchona production in Mysore. However, it was in South America—the only tropical continent in the world that had previously been malaria-free and that, until now, had profited least from quinine's extraction—that the new control measures were about to enjoy their finest hour.

EVER SINCE THE SPANISH CONQUISTADOR Vasco Núñez de Balboa first laid eyes on the "southern sea" from a hill on the Isthmus of Panama in 1513, men had dreamed of a canal linking the Atlantic with the Pacific. Initially, governments had thought it a simple matter of digging and dredging. But as Joseph de Jussieu could have told them, the isthmus was a death trap: in 1735 the French botanist had almost succumbed to fever at Puerto Bello, reviving only after taking an infusion of cinchona bark. And La Condamine and the other academicians had been so anxious to flee Puerto Bello's "dangerous climate" that they had hurried on to Ecuador after only a few days.

The *Anopheles* was only one of the reasons the isthmus was so feared. Panama also teemed with *Aedes aegypti*—the carrier of yellow fever—and in combination, the two vectors were deadly. If the "yellow jack" didn't get you in the first week, malaria inevitably would.

The French learned the lesson the hard way. In 1879, Ferdinand de Lesseps, triumphant from engineering the Suez Canal, began laying a railway line across the isthmus in preparation for gouging a canal through the jungle. No one knows exactly how many Frenchmen died—twenty-two thousand was one estimate—or what proportion of the fatalities were black laborers imported from the West Indies. But it is said that for every sleeper laid, a man lost his life, and in 1889 Lesseps was forced to abandon the project.

It took a man in the Rossian mold to realize the scale of the problem and draw up an appropriate battle plan. That man was William Crawford Gorgas, a colonel in the American medical corps who had been raised in the malarious state of Virginia. Gorgas was not an instant convert to Ross's theories. He came to mosquito control reluctantly, through his experiences with yellow fever, first at Fort Brown, Texas, where in 1882 he was stricken with the disease and almost died, and later in Havana, where he came under the sway of the Cuban doctor Carlos Finlay and the U.S. Army surgeon Walter Reed, a brilliant medical researcher who had previously studied the typhoid bacillus.

Since 1881, Finlay had been arguing that yellow fever was transmitted by "stegs"—the *Aedes* mosquito, also known as *Stegomyia*—but it was only when Reed was sent to Havana in 1900 to investigate the disease and allowed *Aedes* to feed on three of his assistants that Finlay's suspicions were confirmed (they all died). Although Gorgas remained "unconvinced" by these experiments, he immediately instituted a strict quarantine and set about oiling the cisterns and pools in Havana's backstreets where the mosquitoes bred. By the summer of 1902, yel-

Postcards by Albert Guillaume, issued by the French War Ministry during World War I. (Wellcome Library, London)

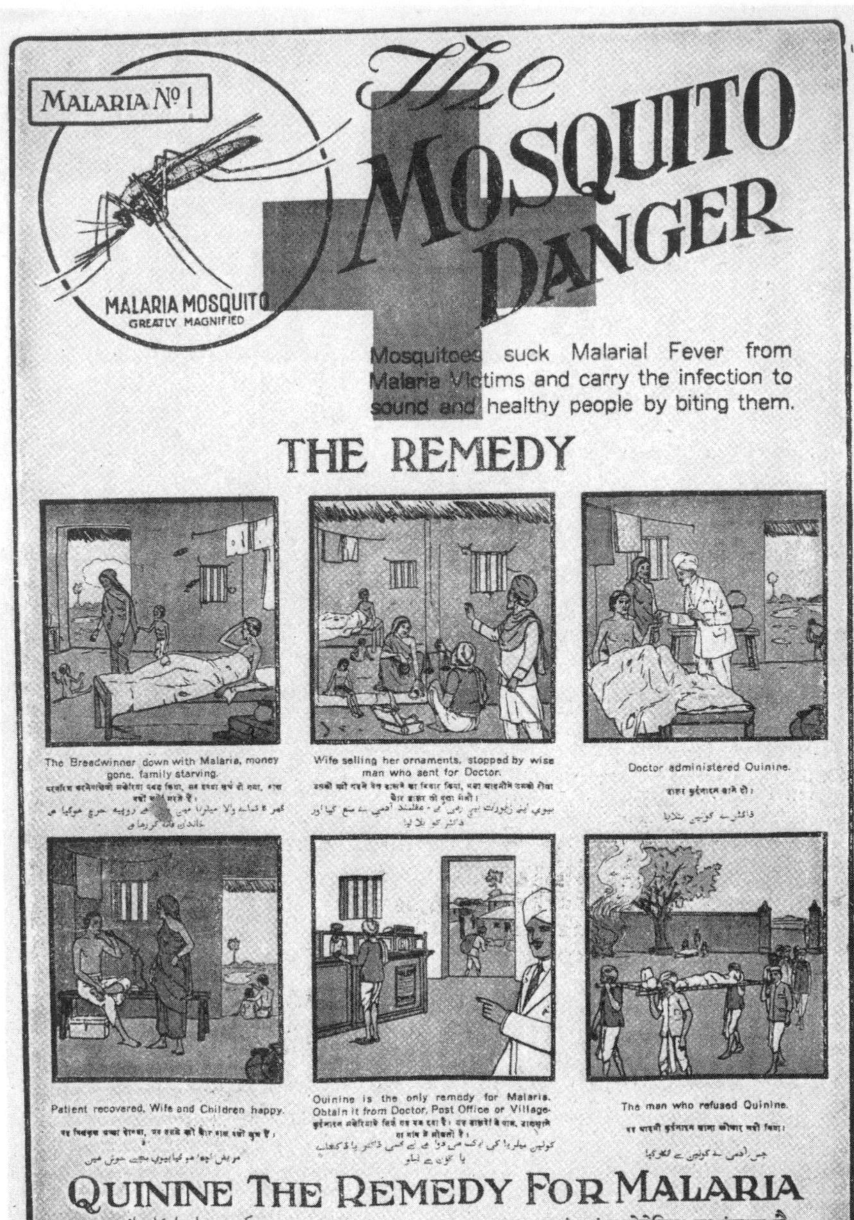

A 1927 poster from British India spelling out the benefits of quinine and the dangers of not taking it. (Wellcome Library, London)

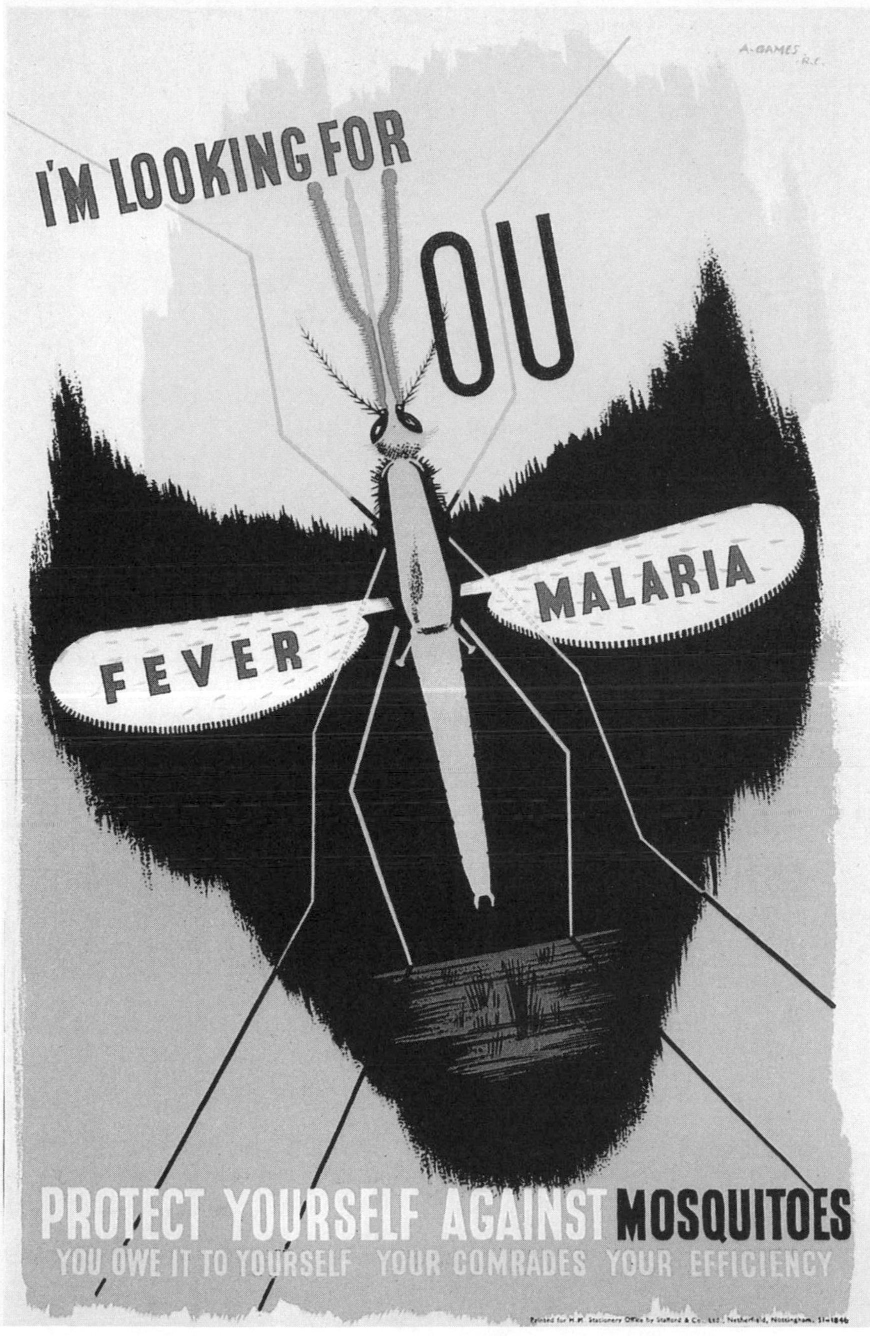

A British public health poster warning of the dangers of mosquitoes and malaria.

(Wellcome Library, London)

A Royal Air Force cartoon showing the three clinical stages of a malaria attack.

The Roman Campagna Malaria Commission carrying their gear. (Wellcome Library, London)

An advertisement for Burroughs Wellcome's "tabloid" medicine chest.
(Wellcome Library, London)

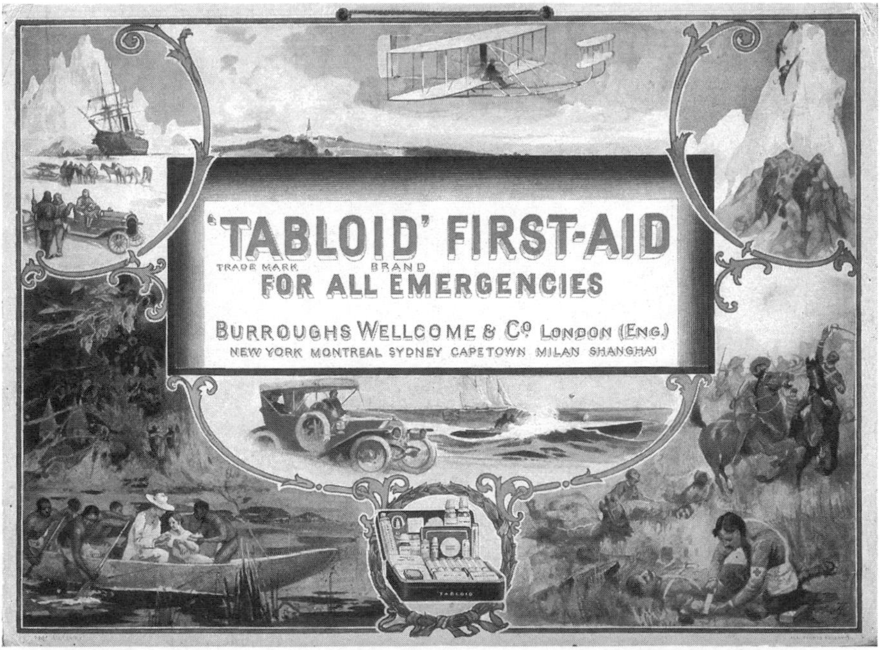

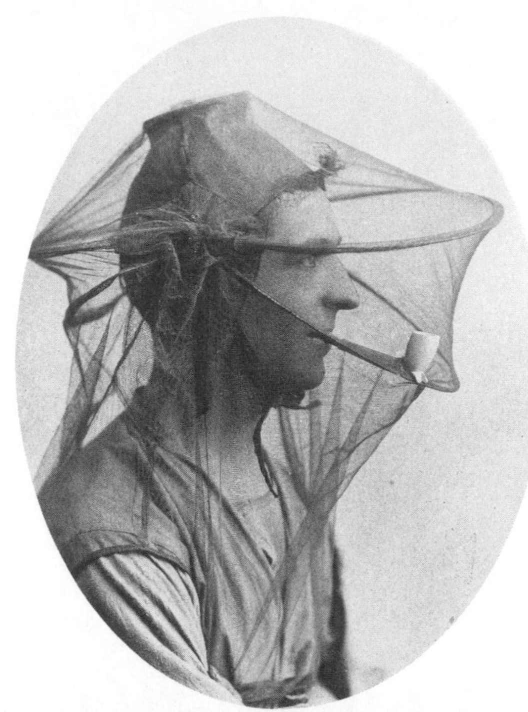

Pipe dream: a British malariologist poses in essential tropical headgear.
(Wellcome Library, London)

Below: Cinchona bark being dried in the sun, circa 1950.
(Copyright © Hulton-Deutsch Collection/Corbis)

A cartoon celebrating Charles-Louis-Alfonse Laveran's discovery of the malaria parasite in 1880.

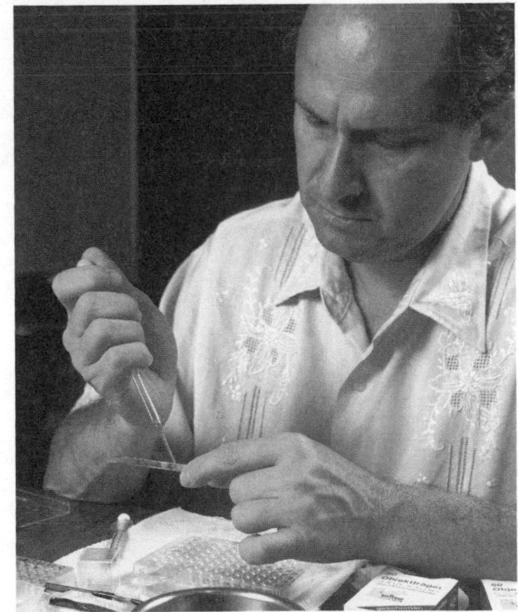

Manuel Elkin Patarroyo, the Colombian immunologist who hopes to develop a malaria vaccine.

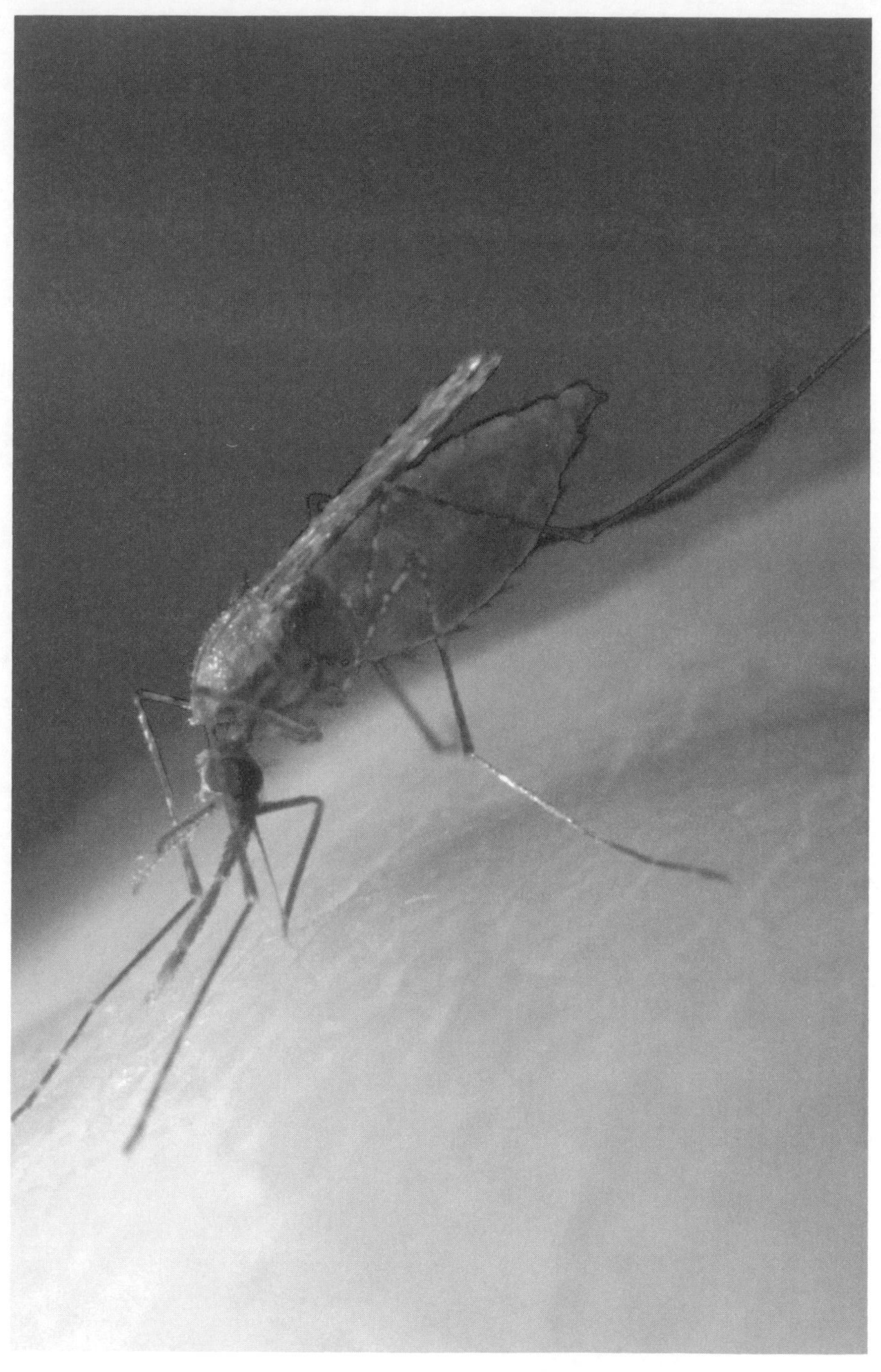

Anopheles gambiae, the deadliest insect in the world. (Science Photo Library)

low fever had been completely eradicated, and deaths from malaria had been reduced to seventy-seven. "For the first time since the English occupation in 1762," Gorgas wrote to Ross, "we have an October free from yellow fever, and malaria decreased more than one half."

When the United States began taking an active interest in the problem of the canal route in 1904, Gorgas was the natural choice to head the campaign against the mosquito. He initially encountered stubborn opposition from Canal Zone Commission officials who considered the mosquito-transmission theory "balderdash," but the chief engineer, John Stevens, had seen workers suffering firsthand and decided to give him a chance. Getting rid of yellow fever was relatively straightforward. As in Cuba, Gorgas launched an all-out attack on *Aedes*, screening the water barrels where they bred and coating ponds, cisterns, and cesspools with oil once a week. At the same time, he launched an extensive pyrethrum fumigation campaign and, as a final precaution, screened in those who had already contracted the disease. By 1906 yellow fever had been completely eradicated.

Reducing malaria took a little longer and required four lines of attack. First, Gorgas eliminated the *Anopheles*' breeding grounds by draining and filling in swamps and spraying oil on lakes, streams, and ditches the length of the canal route. Second, he employed full-time "swatters" to kill mosquitoes harbored in workers' quarters. Third, he erected screens outside every veranda. Fourth and last, he deployed quinine. Establishing a chain of more than twenty dispensaries between Cristóbal and Balboa, he offered men a prophylactic dose of 150 mg twice a day. In one year alone, these makeshift health centers dispensed more than one ton of quinine. To get around some men's aversion to its bitter taste, Gorgas even offered the drug diluted in "extra-sweet pink lemonade."[4]

Within three years Gorgas had cut the incidence of malaria by more than half, with a steady decline in cases thereafter, from eight hundred out of every thousand canal workers in 1906 to seventy in every thousand by 1913. And by the time the first ship passed through the canal in 1914, deaths from all causes had fallen to just six per thousand. It would not be long, Gorgas predicted, before "life in the tropics for the Anglo-Saxon will be more healthful than in the temperate zone."

The construction of the Panama Canal was a remarkable achievement, not least because it showed that disease need no longer be considered a barrier to progress. The lessons were clear. If you were determined enough and brought military organization to bear, the tropics could be tamed. Well, almost. By 1916, Gorgas's measures had succeeded in reducing malaria cases in Panama to sixteen in every thousand, but, unlike with yellow fever, he would never completely eradicate the disease. The reason was that the workers continued to harbor the gametocytes in their blood, and until every last *Anopheles* in Panama was exterminated—a truly impossible undertaking—malaria could rebound at any time.

IT IS NOT KNOWN what Markham made of the canal. By the turn of the century his energy and attention were focused on another geographical challenge—mapping the Antarctic. In 1901 he had dispatched Robert Falcon Scott to the South Pole on the *Discovery*, and by 1910 he was fully occupied with fitting out Scott's second, tragic expedition. However, Markham had always been fascinated with the problem of the circumnavigation of the globe, and the linking of the seas must have struck him as a remarkable achievement, not least because during the cinchona mission he had had to cross the isthmus twice—once on the way out from Southampton, the second time on his way back with the seeds. Through his histories of Columbus and the early Spanish colonists, Markham was also well aware of the isthmus' dreadful reputation for fevers. Now, thanks in part to his humanitarian efforts, quinine had helped make the isthmus safe.

Unfortunately, the same was not true of India. The failure of the Mian Mir project coincided with a serious epidemic. In 1911 malaria raged throughout the country with a severity comparable to the epidemic ten years earlier. The problem was not the lack of quinine but the authorities' failure to mount a consistent, military-style campaign against the *Anopheles'* breeding grounds. Irrigation and drainage works had been commenced, but they had not gone nearly far enough. Florence Nightingale's prescription fifty years earlier that "the salvation of the Indian Army must be brought about by sanitary measures everywhere" had been ignored. Regular malaria surveys were commissioned but they just confirmed what everyone already knew. As the Malaria Advisory Board put it, the war against malaria required "continuity of policy and extreme thoroughness." Without both, the parasite would never be defeated. Nevertheless, Markham could at least take comfort in the fact that the cinchona missions had brought the cure within the reach of ordinary Indians. Or could he?

In June 1911, Leeds University had presented Markham with a doctorate in science. Conferring the degree, the university's chancellor, the Duke of Devonshire, had singled out his missions for unique praise: "To have established in our Indian Empire . . . the cultivation of a prophylactic for its desolating malarial disease, was a service to humanity such as few may hope to render by the undivided labours of a lifetime."

But Markham's philanthropic vision was already in ruins. The problem was that despite the evidence that quinetum was as effective as quinine at curing malaria, the medical profession preferred the refined drug. By 1913, consumption of Markham's cheap cinchona febrifuge had fallen to just sixteen hundred pounds, and India—in common with the rest of the world—was receiving most of its quinine from Java. In 1910, Markham had written to the secretary of state for India "deploring" the rundown state of the Indian cinchona plantations, but rather than encourage planters in Ceylon to switch to *Cinchona ledgeriana* production, in order to compete with the Dutch, the British simply leased their own plantations in Java. At first the arrangement worked well. After all, it mattered little where the bark came from as long as the end product—quinine—was plentiful and cheap. Indeed, rivalry between the planters in Java and the European manufacturers who controlled the production of quinine had generally exerted a downward pressure on quinine prices. In 1892, for instance, in an effort to keep bark prices low, the manufacturers had formed a syndicate under the direction of the German quinine producers to ensure that no one outbid the others. But the Dutch planters resented the European manufacturers' monopoly, and in 1896 they set up their own factory in Bandoeng, Java, with the aim of manufacturing quinine directly for the Amsterdam market.[5] By 1898 the Bandoeng factory, producing some 220 pounds of quinine a day, began undercutting European quinine prices. Paradoxically, the reduction of bark shipments to Europe brought about by the Bandoeng factory led to a steady rise in the unit price of bark on the Amsterdam market. The result was that bark prices, having reached a low of just over 2 cents a unit at the end of 1896, began to rally again, rising 6 cents in the first eleven months of 1897, and peaking at 12.25 cents a unit in September 1900.[6]

By 1911, the European manufacturers had come to the realization that the only way to protect their profits was to invite the Dutch planters to join their accord. Two years later, in 1913, this is exactly what happened. Under an agreement known as the First Quinine Convention, the European manufacturers agreed to purchase the bark equivalent of 512 tons of quinine annually from the Javanese planters at the minimum price of 5 Dutch cents per unit. If quinine prices increased over a certain fixed amount, the manufacturers would share their excess profits with the planters by increasing the amount they paid for a unit of bark. To ensure that no one cheated by, for instance, exporting excess bark to third parties or dumping quinine, a committee, consisting of three planters and three manufacturers' representatives, was established in Amsterdam to enforce the agreement. Its name was the Kina Bureau.

The idea was that the bureau would regulate the agreement for five years, from July 1913 to July 1918. Clearly, it was a monopolistic arrangement. The only thing acting as a brake on prices was the excess supply of bark from the Java plantations. If, however, demand for quinine should suddenly go up or the plantations should suddenly contract, there would be nothing to prevent the bureau setting any price it liked. Before the Kina Bureau had a chance to flex its muscles, however, the First World War broke out and the British and French manufacturers were forced to sever ties with their German confreres. Within a year, Allied troops were dug in all along the Eastern Front with only the drone of mosquitoes for company. The war was to change everything.

11

The Malaria Fighters

> The history of malaria in war might almost be taken to be the history of war itself.
>
> —C. H. MELVILLE

JUST NORTH OF WASHINGTON, D.C., is a monument to the military's enduring fascination with malaria. From the outside, the joint U.S. Army–Navy medical research institute looks like any other recent addition to the American suburbs. A half-million-square-foot glass-and-brick box, it rises above the colonial-style houses that line the pretty backstreets of Bethesda, Maryland, like some sinister and unwelcome corporate invader. But once you step across the threshold of "Building 503"—as it is known to the military scientists who labor inside it—the impression changes. Ten huge slate murals line the atrium, softening the light pouring through the plate-glass windows and guiding the visitor into the heart of a state-of-the-art medical research facility. The murals are inset with portraits of famous army doctors, like Walter Reed—the man who put the mosquito in Gorgas's ear, so to speak, and helped eliminate yellow fever from Cuba. But it is not the images of the mustached pioneers of military science that make the most lasting impression, but the imprecations against malaria etched in the slates. "This will be a long war if for every division I have facing the enemy, I must count on a second division in hospital with malaria and a third division convalescing from this debilitating disease," reads one from General Douglas MacArthur in 1943. And from C. H. Melville in 1895: "The history of malaria in war might almost be taken to be the history of war itself."

Melville was exaggerating, but only slightly. Malaria has rarely been the decisive factor in wars. But from the death of Alaric the Goth from fever at the gates of Rome, to Cartagena, Walcheren, and the American Civil War, it—or at least the mosquito—has repeatedly influenced the outcome of military engagements. During World War II, for instance, nearly half a million American servicemen were stricken with malaria, and in the southwest Pacific the enforcement of malaria prophylaxis was a key factor in the eventual Allied victory. The relationship is far from one-way: just as malaria has affected warfare, so the legacy of war has often been an increase in malaria. In 1923–26 there were ten million cases and sixty thousand deaths following the devastation wrought by the civil war in Russia, and it was during the Vietnam War that drug-resistant strains of malaria appeared for the first time in Southeast Asia. That is why army doctors have a special responsibility for malaria and why the language of malaria control is so often couched in military terms. From Ross's "mosquito brigades" to Gorgas's "war on the mosquito" to the World Health Organization's benighted "global eradication" campaign in the 1960s, we have imagined that it is simply a matter of "blitzing" the mosquito until the scourge of malaria disappears.

But of course it is not that simple. As the twentieth century wore on, the mosquito proved far more resilient than Reed, Ross, or Gorgas could ever have suspected. The WHO no longer talks of eradicating the mosquito battalions ranged across Africa and the Asian subcontinent with DDT, for instance, but of "rolling them back" through a series of continuous campaigns employing bed nets, malaria prophylaxis, and new-generation pesticides. Nevertheless, the link remains and nowhere more so than at the Walter Reed Army Institute of Research (WRAIR) in Maryland.

As pharmaceutical companies have retreated from the front line of malaria research, the heirs to Walter Reed have increasingly had to step into the breach. The U.S. Army invests about $4 million annually in the hunt for new antimalarials. It is pure self-interest, of course. In nearly every engagement in the last hundred years, the U.S. Army and Navy have suffered more casualties because of diseases than because of enemy fire. Nonetheless, without Walter Reed's continuous drug development program, there would be no new antimalarials waiting in reserve and thus nothing to stop the relentless march of drug-resistant parasites across the face of the globe.

MORE THAN ANY OTHER WAR in the last century, it was World War I that convinced Allied military planners that they would have to take an active interest in malaria prophylaxis, and it was in Greece that the lesson was driven home.

There are five species of *Anopheles* in Macedonia, each one specialized in a different type of terrain, but if you are considering a war in northern Greece, there are only two you really need to worry about. *A. maculipennis* is a "sea level" mosquito. It is found all over Macedonia, but it is most prevalent in the Struma and Vardar valleys and the deltas of the Vardar and Galiko rivers west of Salonika. It breeds in lakes, marshes, the holes left by horses' hooves, and even in discarded tin cans. Its cousin *A. superpictus* rules the hill country. It breeds in brackish water and pools at three thousand to four thousand feet above sea level, but it is also found lower down the valleys, near the point where the streams break upon the plains. From December to March it is so cold in Macedonia that neither species has much opportunity for reproduction, but in May the temperatures—especially in the valleys—begin to soar and by August, Salonika is a sweltering 90 degrees Fahrenheit.

During World War I, Britain and France at first hesitated to commit troops to the Balkans, preferring to leave the Serbs to resist the Central Powers alone. But when in September 1915 the Bulgarians threatened to join the assault on the Serb line, the Allies had no choice but to respond. The Allied plan was to evacuate Serb fighters from Yugoslavia and establish a new base with which to block the German advance in northern Greece. The site that the British and French chose was Salonika.

The Franco-British expeditionary forces knew they would face a problem with malaria, which is why they timed their arrival in the delta for October—the end of the mosquito-breeding season. What they could not have known was that hundreds of thousands of Greek refugees infected with falciparum malaria had fled to Macedonia from Asia and the Caucasus just before the landings, greatly stoking the levels of infection.

The Allies' first mistake was to set up camps and field hospitals in the hill country above Lake Langaza, in the belief that they would be out of range of *Anopheles* come the summer breeding season. Their second was to advance up the Struma and Vardar river valleys the following June, seemingly oblivious to the *A. maculipennis* multiplying in the marshes.

By July the British and French were strung out in a long line, running east–west from the mouth of the Struma River to Lake Dojran and over the hills to Vardar and the Monastir Road. The result was catastrophe. Hospital admissions for malaria rose steadily, from 4,500 in July to 7,500 in August and close to 9,000 in September 1916. The following summer was even worse, with 30,000 cases in September and October alone.

Ross, by now a lieutenant colonel, visited the Monastir Road and suggested forming mosquito brigades to drain marshes, clear streams, and kill larvae. In addition, bivouacs were improvised and special bell-shaped nets manufactured. But nothing seemed to work. By the end of 1917 a further 70,000 troops had come down with malaria and the field hospitals were so overwhelmed that medics were having to evacuate the most chronic cases to England at the rate of 1,000 a week.

The final figures make sober reading. By October 1918, out of a force of 160,000, there had been 162,000 hospital admissions, half of them for falciparum malaria. In other words, some men contracted malaria more than once. Nor were the consequences confined to Greece. Men returning from the front brought the disease home, sparking small outbreaks in Kent, and long after the war was over many veterans continued to suffer attacks.

SALONIKA UNDERLINED the importance in warfare not only of malaria but also of quinine. Every morning troops were expected to fall in in an orderly line and take their medication. Although officers complained that a quinine prophylaxis regimen was difficult to enforce—both because of the drug's unpleasant side effects and the old soldier's canard that it caused impotence—there was little doubt about the drug's value. Despite the huge numbers of casualties at Salonika, there were just 821 deaths from malaria during the whole campaign. Quinine may not have prevented soldiers' contracting the disease, but once they had been admitted to hospital, it saved their lives. And it wasn't just in Macedonia that quinine was needed. In East Africa, malaria had decimated Indian troops specially selected for colonial duty in the mistaken belief that they would be immune to it. Similarly, when the British pursued the Turks into the malarious Jordan Valley in the autumn of 1918, they were confronted by propaganda leaflets warning, "The flies die in Jericho in July, and the men in August." It was no idle threat: within weeks almost half the 40,000-strong British force had been hospitalized with malaria.

Soon the British army medical department was clamoring for quinine, but because of the neglect of the Indian plantations there was now only one place to get it—Java. At the outbreak of hostilities, the Quinine Convention had been torn up and both Britain and Germany had imposed export controls on the drug. The war was causing severe disruption to shipping lanes, and it was feared that with supplies of bark from Java becoming increasingly erratic, quinine stocks would rapidly be exhausted. In 1917, in an effort to cut off quinine to the Central Powers, the British forced the Dutch to close the German-owned Chininefabriek factory in Amsterdam. However, Germany had taken the precaution of cultivating hybrids of *Cinchona ledgeriana* and *C. succirubra* in East Africa, and by 1917 its Tanzanian plantations were producing six thousand pounds of top-quality bark a year. Fortunately, in early 1918 the British quinine manufacturer Howard and Sons was able to reach an agreement with the six British-owned plantations in Java to buy up their entire annual output of bark. Then, in September, the Allies also agreed to buy the entire output of the Bandoeng factory, insisting on an immediate delivery of half a ton of quinine and 400 tons of bark. By the end of the war, Bandoeng accounted for more than three quarters of the world's production, and quinine had once again become a priceless commodity.

It was a sad end to Markham's philanthropic dream, but fortunately he was spared the anguish of seeing it. In 1905 he had stepped down from the Royal Geographical Society after twelve years' service as president—the longest term of anyone before or since—and in 1912 he resigned from his other positions. Now, aged eighty-two, his aim was to write one last history—a compendium of everything that was known about "the Ends of the Earth," the polar regions to which he had devoted much of his life, both as an explorer in Greenland in 1875, and more recently as the sponsor of Scott's two voyages to the Antarctic. Then, in February 1913, came the news that Scott had reached the South Pole the previous year only to perish in a blizzard on the return journey. Markham had loved Scott like a son and was deeply saddened by his death. Perhaps that explains why, when a deputation of old boys from Westminster asked him to sit for a portrait for his alma mater, he chose to allude not to the voyages of the *Discovery* or to his own polar adventures but to an earlier epic—the hunt for red cinchona. Markham was presented with the painting, showing him seated beside Pavon's *Cinchona uritusigna*, by the artist George Henry at a gathering of friends and well-wishers in Cambridge. The painting's only reference to Markham's "other" career, as the father of polar exploration, was a small silver sleigh perched on the mantelpiece.

The following year, despite the war raging in western Europe, Markham traveled with Minna to Madeira for a holiday, dodging German submarines on their return via Portugal. By January 1916, he was very much preoccupied with the war, confiding in his diary that it was "a very sad and portentous New Year . . . [with] at home the enemies of efficient methods of resistance still clinging to office." But although he recorded that Belgium, northern France, Poland, and Serbia were "all conquered," there was no mention of Salonika or of the mounting importance of quinine to the war effort. Instead, almost his last thoughts were connected with the death of Scott.

"Sturdy little Peter Scott came and walked with us in the Square garden," reads the last entry but one in his diary. "I often think of his dear father and the men he has trained to fight his country's battles."

A few days later, on January 29, he was reading in bed by candlelight at his home in Eccleston Square when the candle toppled and set fire to his bedclothes. Although he screamed for help and the flames were quickly extinguished, it seems that the shock was too great for him. After lying unconscious for twenty hours, he died. He was eighty-five.

Why he chose to read by candle when there was a perfectly good electric light in his room is not known. Perhaps it was a habit he had picked up in the navy and, despite having swapped his hammock for a bed, was unable to change. It was a gruesome and—one is tempted to say—cursed end, almost as if Markham were driven to his grave by the same flames that had consumed Ruiz and Pavon's collections in Peru two centuries before.

BY THE END OF THE WAR the price of quinine had more than doubled, and by 1920 it had reached a high of 52 florins a kilo. But peace gradually brought a return to economic normality. Dutch planters could no longer count on the military to buy up all their excess bark. Furthermore, with the worldwide decrease in purchasing power that followed the war, neither could consumers be relied on to maintain quinine's price. Some way had to be found to grow sufficient bark to fulfill the requirements of European quinine manufacturers, but at a price that would ensure that planters had an incentive to remain in business. The result was the Second Quinine Convention.

Under the new agreement, signed in 1918, manufacturers agreed to pay planters a minimum of 6 Dutch cents per unit for their bark—one cent more than the prewar price—and guaranteed they would buy 748 tons in the first year of the contract and 512 tons in each year thereafter. In addition, if the price of quinine exceeded the contract price, then the manufacturers would split the difference with the planters by paying higher unit prices for their bark.

Overall it was a good deal for the manufacturers and an excellent one for the planters. The problem from consumers' point of view was that the only signatories were the Dutch—the European manufacturers having gone out of business during the war, leaving just the three Netherlands factories and the one in Bandoeng. Eighty percent of the world's production of quinine was now regulated by the Kina Bureau in Amsterdam. In theory, the bureau could set any price for quinine it wanted and the world would have to pay. Although by 1924 the price of quinine had fallen to 35 florins a kilo, the U.S. government argued that it was still too high and that quinine could be produced far more cheaply from barks sold at cost price. As a statement of bald economic fact this was undeniably true. There was just one flaw in the argument: if you reduced the price of the bark too far, no one could be persuaded to grow it.[1] The truth was that the limits of "poor man's quinine," as one postwar writer put it, had already been reached in India, where the British had pioneered a cheap febrifuge—quinetum—only to find that at cost price it was still too expensive for the vast majority of Indians. By 1921, only a sixth of malaria sufferers in India had access to quinine, and deaths from the disease were running at 1.3 million annually. One commentator estimated that just to treat those attending hospital, the government would have to import 56 tons of quinine a year but that the true demand could be closer to 674 tons.[2]

In 1917, Herbert Hoover, then an up-and-coming food administrator in the U.S. government, had been so alarmed by the prospect of a wartime quinine shortage that he had telegrammed U.S. officials in the Philippines with just two words: "Raise quinine." Now, ten years later, the U.S. Department of Justice decided to launch its own war, bringing an antitrust suit against the Kina Bureau, accusing it of price-fixing. In 1928 federal agents impounded five tons of Dutch-produced quinine at the port of New York. At the same time, the Department of Commerce, acting on Hoover's advice, began encouraging U.S. quinine manufacturers to invest in their own plantations so as to be independent of Java for supplies of bark. Before the plantations could become established, however, the United States reached an accord with the Kina Bureau under which the bureau agreed to refrain from further monopolistic practices.[3]

By now the limitations of quinine were apparent, not only to the military but also to economists. What the generals wanted was a drug that could be given to the troops before they entered a malarial area and that would ensure they never developed the disease. What the economists and strategic planners wanted was an antimalarial that was cheap and whose production could be turned on and off at will without having to worry about supplies falling into enemy hands. The problem with quinine was that it was short-acting, and toxic if maintained at the levels needed for a prophylactic dosage. Common complaints included nausea, blurred vision, and ringing in the ears—all of which made a medical regimen difficult to enforce. After nearly three hundred years of continual use, the miraculous "fever bark" had run its course. In the future, the solution to malaria would no longer be sought in nature but in synthetics, or so the military thought.

EVER SINCE PELLETIER AND CAVENTOU isolated quinine from cinchona bark in 1820, chemists searched for a way to synthesize the drug. The first to attempt it was the German chemist Friedlieb Runge in 1834. But though Runge succeeded in deriving quinoline, an organic base, from coal tar, the synthesis of quinine defeated him. In 1856 a young British chemist, William Perkin, was experimenting with quinine synthesis when he stumbled on mauveine, the first of the aniline dyes. Perkin quickly abandoned the hunt for a quinine substitute and devoted himself instead to marketing his lucrative new dyestuff.[4]

Curiously, it was while experimenting with another synthetic dye that a scientist made the first significant advance. In 1891 the German immunologist Paul Ehrlich used methylene blue to treat two sailors who had returned to Berlin with tertian malaria. Ehrlich had already noted how methylene blue would stain—combine with—blood cells and certain microbes. Now, to his surprise, he saw that the dye also had a powerful inhibiting effect on the parasites in the sailors' blood.

However, because there was no suitable model at the time for comparing human and animal malarias, Ehrlich did not pursue the research further. Instead, he began experimenting with other dyes in his search for a "magic bullet" for syphilis, synthesizing numerous compounds until, in 1909, on his six hundred and sixth attempt, he came up with arsphenamine (Salvarsan).

Indeed, it was only when scientists confirmed in the 1930s that the life cycle of the malaria parasite in birds was identical to falciparum infections in humans that it became possible to investigate synthetic therapies for malaria in a systematic way. Again, the Germans led the way, and once again World War I was the impetus.

Because of the Allied blockade on quinine from Java, German troops in East Africa had suffered even more than the British. Determined never to be a hostage to quinine again, the German government asked its scientists to renew the search for a substitute. The center of operations was IG Farben, part of the Bayer Dye Works in Elberfeld. (IG Farben was later notorious for its dealings with the Nazi government.) Beginning with methylene blue, they started testing thousands of compounds until, in 1926, they hit on one that worked. Its name was pamaquine (Plasmochin), and it possessed a distinct advantage over quinine: it rendered the sexual stage of the parasite noninfectious. Unfortunately, it was also highly toxic and was soon abandoned. Then, in 1932, IG Farben scientists discovered quinacrine (Atabrine), a dye almost identical in its action to quinine. In fact, although it turned patients' skin yellow, in many ways it was superior, persisting in the blood for at least a week after the first dose and providing better long-term prophylaxis. Once again, however, it was toxic and difficult to tolerate.

The breakthrough came in 1934 with the synthesis of Resochin. A few years later Sontochin was synthesized. These were a new type of synthetic—a class of compounds known as four-amino quinolines. Like quinine, they contained a benzene ring fused to a pyridine ring. But instead of quinine's complex alcohol and vinyl side groups, the German scientists substituted simple side chains. The resulting compounds targeted the blood stage of the parasite, like quinine, but were much faster acting. However, IG Farben scientists exaggerated the compounds' toxic side effects and did not go on to explore the 4-quinoline group further, possibly, it is said, because Hermann Göring held the patent and would have been the only one to have profited from sales. This was a big mistake, because as U.S. scientists at IG Farben's sister company, Winthrop Stearns, would discover ten years later, both compounds were closely related to chloroquine, possibly the best antimalarial ever developed—indeed, Resochin and chloroquine were subsequently found to be identical. Ironically, IG Farben had passed the formula for Resochin to the U.S. company in the late 1930s, but Winthrop simply shelved it. What persuaded them to take another look was the outbreak of World War II.

This time around it was the Allies who were deprived of quinine. With the German invasion of Holland, stocks at the Kina Bureau's plants in Amsterdam and Maarsen became the property of the Nazis.

Then, in 1942, the Japanese invaded Java, seizing the Dutch plantations and with them 90 percent of the world's supply of bark. The U.S. War Production Board issued an emergency decree restricting the use of quinine to the military and launched a desperate campaign to pool chemists' dwindling stocks. To aid the effort, President Roosevelt even forwarded one hundred pounds of the drug that he had been given as a gift by the president of Peru. The U.S. Board of Economic Warfare began looking for alternative sources of bark. Deals were quickly struck with the Andean republics to buy up their annual output. But because of the success of the Java plantations, very little bark was now under cultivation. Then there was the problem of quality control.

Like the seventeenth-century European apothecaries before them, American chemists now discovered that the quinine content of the wild South American barks was extremely variable. There was no alternative, decided the planners and military strategists, but to recruit a group of botanists prepared to risk their lives by returning to the bark forests to gather seeds and plants of the most valuable species. Beginning in 1942, this is exactly what happened: the Board of Economic Warfare organized a series of missions to Colombia, Ecuador, and Peru manned by handpicked botanists and forestry experts from the Department of Agriculture and leading U.S. universities. Their job was to follow in the footsteps of La Condamine, Mutis, Spruce, and Markham, exploring some of the most inaccessible parts of the Andes in search of species with obscure names like *Cinchona pitayensis*, *officinalis*, *succirubra*, and, of course, *calisaya*.

Travel in the Andes had changed little in a hundred years, and the botanists faced many of the same risks as did the nineteenth-century pioneers before them: precipitous slopes, recalcitrant mules, roaring mountain streams, and exhaustion from the dizzying altitude. As Froelich Rainey, the director of the Ecuadorian mission, later admitted: "Scrambling up a narrow gorge where mud and water, belly deep, alternated with irregular boulders and rock slides, I learned that none of my former experience with mountain trails and riding animals applied in the Andes." But at least Rainey survived. One of his colleagues, Arthur Featherstonehaugh, a former marine, was not so lucky. Gazing out over the Quitonian Andes at ten thousand feet, Rainey was appalled to see "Feather's" face turn a blotchy gray. "He was panting with such force that he could scarcely speak. After a time he was ready to go on, but he looked desperately ill and we urged him to ride again and rest . . . We were rigging a stretcher for Feather when I heard his low, terrified voice, 'Frol, come here!' Those were his last words . . ."

With the help of Ecuadorian botanists, Rainey eventually found Spruce's red bark growing in the foothills of the eastern Andes near El Topo. The result was extensive new Ecuadorian plantations, producing totaquinine. Similar plantations with local species were also begun in Peru and Colombia, and by the end of the war South America was contributing twenty-six hundred tons of bark a year to the Allied war effort. The only country that refused to cooperate was Bolivia. Instead, like the Dutch a hundred years before, the Board of Economic Warfare was forced to dispatch agents to La Paz to try to locate promising stands. Like Hasskarl, they had little success, and once again *calisaya* remained the most elusive bark of all.

However, there was one place the Americans knew they could obtain *calisaya* seeds—the southwest Pacific. When Herbert Hoover had ordered officials in the Philippines to "raise quinine" in 1917, no one had taken much notice. But four years later the Philippines had a governor-general with the medical background to realize the importance of Hoover's order. General Leonard Wood was a Harvard-educated doctor who had begun his army career in the medical corps. As governor-general in Cuba in 1899, he had funded Walter Reed's yellow fever studies. When he saw a young forest ranger in the Philippines die of malaria in 1921 for lack of quinine, he decided to find the money to carry out Hoover's order. Commandeering $4,000 from a special contingency fund, Wood sent the money to the U.S. consul in Java and received a bottle of seeds of *Cinchona ledgeriana* by return military post (the price had gone up somewhat since Ledger's time). Those seeds became the responsibility of Arthur Fischer, the director of the Philippines Bureau of Forestry. Within six years Fischer was raising *C. ledgeriana* and *C. succirubra* plants at Baguio, on Luzon. Then, in 1927, he moved the plantations to Mindanao, the island at the southern end of the island chain, where the climate was more humid. In 1936 the first shipment of bark from Mindanao reached Manila, and by 1941, Fischer had established a factory for the production of totaquinine in the Philippine capital. By the time the Japanese invaded Pearl Harbor, it was producing 2.5 kg of the febrifuge a day.

THE PHILIPPINES, along with Papua New Guinea and the Solomon Islands, is one of the most malarious places on earth. Before World War II, it is estimated that as many as two million people were infected with the disease, with ten thousand dying each year. The only thing that kept the mortality rate from spiking even higher was quinine, but with the Japanese invasion early in 1942, the American and Filipino troops garrisoned on the island suddenly found themselves cut off from their supply lines. As the Americans retreated to Bataan and Corregidor, infections shot up, and by early spring 80 percent of the troops had come down with malaria. Among them was Arthur Fischer, now a colonel in the intelligence service. In March 1942, ill with fever, he persuaded his commanding officer to allow him to risk the Japanese anti-aircraft batteries and fly to Mindanao, where he hurriedly improvised a makeshift factory for the production of the cinchona febrifuge. But before he had an opportunity to return, the weakened American forces in Bataan surrendered, leaving Fischer and just a small guerrilla force to fight a rearguard action against the Japanese in Mindanao. Armed with totaquinine, the guerrilla fighters maintained their resistance, and according to Father Haggerty, a Catholic priest who remained on Mindanao through the occupation, the bark was shipped all over the southern Philippines, saving "tens of thousands of lives."

Meanwhile, General Douglas MacArthur, the commander of the South Pacific forces, who had ordered the hasty withdrawal to Bataan and who now had himself retreated to Australia, issued orders for the evacuation. Three Flying Fortresses were dispatched to Mindanao with instructions to bring back important military personnel. The first two planes were shot down, but the third, containing Colonel Fischer and a milk churn packed with *C. ledgeriana* seeds, got through. No sooner had he landed in Australia than Fischer and his churn were on the next plane to Washington, where, still feverish, he delivered the seeds to the Department of Agriculture. Within days those seeds had germinated and *ledgeriana* plants were being raised in a government hothouse in Glenn Dale, Maryland, not far from where the joint forces medical research institute stands today. By the end of the war, the plants had been transplanted to Ecuador and were thriving on the higher slopes of the Quitonian Andes. Thus, by this curious and circuitous route, Mamani's *calisayas* returned to the continent from which they had been torn.

Unfortunately, the *ledgeriana* plantations in Ecuador came too late to aid the war effort. Besides, by 1943 the United States had found a better solution to the quinine shortage: Atabrine. Army researchers had first tested the compound on soldiers in Panama in 1938, but discarded it after troops reported unpleasant side effects such as vomiting, diarrhea, and yellowing of the skin. Spurred on by the war, however, the government in 1942 brought army and civilian chemists together under the auspices of the National Research Council to test synthetic compounds, both old and new. At first it was feared that Atabrine might contain a toxic substance not in the original German product, but Winthrop researchers soon confirmed this was not the case, and in the autumn of 1942 the army began distributing the drug to troops in the southwest Pacific.

The most pressing need was in Buna, Papua New Guinea, and on Guadalcanal, in the Solomon Islands. In August 1942, U.S. marines had made an unopposed landing on Guadalcanal following intelligence reports that the Japanese were planning an airfield there that could block sea-lanes between the United States and Australia. The Japanese quickly reinforced the island and brought in their navy, effectively isolating the marines in an area of high falciparum transmission. At first little attention was paid to malaria discipline. As one senior officer reportedly put it: "We are here to kill Japs and to hell with mosquitoes." But within months of the landing, attitudes changed dramatically as the force was ravaged by the disease. By the end of the year, 80 percent of the marines had contracted malaria, and one infantry unit had had to be evacuated to New Zealand for rehabilitation.

On Buna, where American and Australian troops had launched a joint offensive in November, the situation was even worse. General Robert Eichelberger, who had been ordered by MacArthur "to take Buna or not come back alive," found that the two-hundred-man rifle companies he had been sent to relieve were down to sixty-five men, all of them febrile. In total, in the fighting throughout Papua New Guinea, nearly 28,000 Australian and American troops had been admitted to field hospitals with malaria in 1942. In comparison, casualties from the fighting were just 7,700. These sorts of statistics forced MacArthur to take malaria seriously.

At first soldiers rebelled against the new Atabrine discipline. Army medics had yet to refine the dosage, and the side effects were so unpleasant that many men threw the tablets into latrines or buried them under mattresses rather than take them. Further trials were conducted in both the United States and Australia, and after it was established that the drug was better tolerated with a large primer dose followed by smaller maintenance doses, the therapy was renewed in the spring of 1943, this time with success. In Papua New Guinea, in particular, Australian medics showed that if a tablet of Atabrine was taken each day, a soldier could operate for months in an endemic zone, and that if the therapy was continued after the soldier left the area, falciparum was completely eradicated from the blood. Not only that, but Atabrine, unlike quinine, eliminated the sexual forms of the parasite, meaning that soldiers returning from the jungle would no longer be at risk of infecting colleagues awaiting transport to the front line.[5]

However, Atabrine was still far from ideal. The "ideal" antimalarial would be chloroquine, but after screening fourteen thousand compounds, neither Winthrop nor the finest scientific brains in the U.S. Army could find it. Then came the fall of Tunis and, with it, the suggestion by French doctors that the Americans might like to experiment with Sontochin, the IG Farben compound they had been using with some success in North Africa. Winthrop researchers analyzed Sontochin's chemical structure in the United States, made slight adjustments to its side chains to give it an even more potent therapeutic and prophylactic action, and renamed it chloroquine. It was only when they compared chloroquine to the supposedly toxic Resochin, the German compound that had been gathering dust on their shelves for nearly ten years, that they realized the two were identical.

The synthesis of chloroquine came too late to make any difference to the war in the Pacific or to the Allied landings in Sicily—another endemic front, where diseases including malaria accounted for more casualties in the U.S. Seventh and British Eighth armies than enemy fire. But following World War II, chloroquine quickly became the drug of choice throughout the tropics. In Ehrlich's parlance, chloroquine was the "magic bullet," a fast-acting, easy-to-administer prophylactic that afforded complete protection against all forms of malaria. Like quinine, chloroquine is what is known as a blood schizontocide—in other words, it hits the parasite when it develops into a schizont in the red blood cells. But unlike quinine, it also takes out the parasites of the vivax, *ovale*, and *malariae* strains in their sexual stage, blocking transmission. The only flaw in chloroquine's armor is that it has no effect on the gametocytes of falciparum. But that was of little consequence because the military now had reinforcements available in the form of pamaquine and primaquine, the new, easy-to-tolerate variants of plasmochin, which did destroy falciparum's crescents and also prevented relapses from vivax malaria. When the British company Imperial Chemical Industries succeeded in 1944 in synthesizing proguanil (Paludrine in the United States), a new compound, unrelated to the quinoline group, that hit both the blood and liver forms of the falciparum parasite, the military's arsenal was complete. At long last it seemed quinine could be consigned to druggists' wastebaskets.

CHLOROQUINE WAS NOT ONLY a boon to the military; it heralded a new dawn for Indian coolies, Thai rice farmers, and millions of ordinary Africans, too. At last, a malaria-free world seemed like a real possibility.

In the 1930s Mussolini had launched an extensive bonification campaign in the Roman Campagna, ending malaria transmission in the Pontina by 1948. In Brazil, a military-style pyrethrum-spraying campaign had succeeded in exterminating *Anopheles gambiae*, the deadly African vector of malaria that had suddenly appeared uninvited in Natal in 1930, sparking the worst malaria epidemic in South American history. Now, adopting the military vernacular, the World Health Organization began trials aimed at nothing less than the "global eradication" of malaria. Its weapons? Chloroquine and a promising new insecticide, DDT. The WHO cleared malaria from Italy in the immediate postwar period, when the deliberate breaching of dikes by retreating German panzer divisions had threatened to undo the prior attempts at eradication; in the 1950s, it virtually wiped out malaria from Cyprus, Greece, Venezuela, Sri Lanka, and—for a time—large parts of India. But by the 1970s, DDT-resistant strains of mosquitoes had begun to emerge, and following a vociferous campaign by environmentalists concerned about DDT's effect on birds and other wildlife, the United States banned it in 1982.[6]

In contrast, there has never been anything controversial about chloroquine. By the 1960s a bottle of chloroquine was as common in colonial homes as bottles of quinine water had been in India a century earlier. Chloroquinized salt was being distributed by the ton to Brazil, Cambodia, the Philippines, and New Guinea, and the U.S. Administration for International Development bought millions of tablets to give away in Africa and India. Unfortunately, chloroquine's honeymoon was over almost as soon as it began. The reason was the evolution of the parasite. Army medics had long known that if left untreated, falciparum could be as deadly as a concealed enemy sniper, but neither military nor nonmilitary scientists had fully appreciated that the malaria parasite also possessed hidden biological cloaking devices and was skilled at taking evasive action.

The first hint of the emergence of drug-resistant strains of malaria came in South America. In 1910, German engineers returning by Amazonian steamer from northwest Brazil came down with falciparum malaria. The men, who had been laboring on a railway construction project near the frontier with Bolivia, dosed themselves with quinine, but on their return to Hamburg, doctors discovered they still had parasites swimming in their blood. The dosage was increased, but it was only after repeated treatments that they were cured. In 1918, German doctors noticed similar resistance to quinine in troops with vivax malaria returning from the Macedonian front. Then, in 1960, in the Magdalena valley in Colombia, and just across the border in the Lake Maracaibo area of southwest Venezuela, doctors simultaneously reported the first cases of chloroquine resistance. These were the very areas where Spruce and Bonpland had come down with malignant tertian fevers more than a hundred years before. In the case of two seismographers who contracted falciparum in the Magdalena valley, not only did the infection not respond to chloroquine, it did not respond to quinine either. One possible reason for the emergence of cross-resistant strains in these areas, scientists theorized, was their proximity to the cinchona forests of Colombia and northern Peru—the places where quinine had been in use for the longest period and where the parasite would have had sufficient time to develop immunity. But in 1962, when scientists also began to see the emergence of chloroquine resistance in Malaya, Cambodia, Vietnam, and Thailand, this theory began to seem less probable. Perhaps resistance was simply the result of underdosage and overdependence on the quinoline-based drugs? Whatever the reason, drug resistance spread rapidly throughout Southeast Asia, and, once again, it was the military that had to find ways to combat the problem.

At the beginning of the war in Indochina, the French were using quinine and Atabrine as prophylactics, but by 1952 chloroquine had become the drug of choice because it was easy to administer—you only had to take one 300-mg tablet a week—and because it had fewer side effects.[7] Chloroquine prophylaxis led to a dramatic decline in infections in the French forces, but no sooner had the United States entered the conflict than the first signs of parasite resistance began to emerge. Army medics initially persevered with the so-called CP tablet—a combination of chloroquine and primaquine that had been shown to be effective against both falciparum and relapses from vivax malaria. By 1966, however, it was clear that the therapy was no longer working. In December 1965, the overall rate of infections in the Army was running at ninety-eight per thousand, and in the Ia Drang valley some units were suffering infection rates as high as six hundred per thousand. There was no choice but to resort to quinine again.

However, except where it was given intravenously, the relapse rates remained high. The medics called up reinforcements—pyrimethamine, a new synthetic that targeted the parasite's enzymes, and dapsone, a slow-acting blood schizontocide—and the infections began to fall. But by now it was apparent that no single drug—whether a synthetic or a purified alkaloid like quinine—could hope to defeat resistant strains of malaria. Whatever chemical you threw at the parasite, sooner or later it would find a way to evade it. What was more, there was now nothing left in reserve.

Gradually, it was beginning to dawn on malariologists that the only form of chemotherapy that could stand a chance of working in the long term was one that combined several chemicals or alkaloids at once—in other words, a combination not unlike the cheap cinchona febrifuge pioneered by Markham in India. Once again, the U.S. government looked to the bark, restoring overgrown plantations in Guatemala, Indonesia, and Zaire and manufacturing totaquinine for distribution to GIs. Thus, three hundred years after the Jesuits brought the "miraculous cure" to Europe, it was still saving lives—this time in Vietnam and Cambodia.

As the war in Vietnam was coming to an end, Wilbur Milhous, a bright young field medic from South Carolina, was called in by his commanding officer. Instead of going to the front, the officer said, how would he like to go back to university and learn something about malaria? "Well, I didn't know the first thing about malaria," recalls Milhous, "but in the army when an officer suggests you do something, you do it." Sitting in his office in Building 503 today, surrounded by souvenirs from the Amazon, Africa, and other endemic malaria regions of the world, Milhous has little doubt he made the right decision.

After getting his Ph.D., he joined the Walter Reed Army Institute of Research in 1983, quickly rising to the rank of colonel. He is now director of the institute's division for experimental therapeutics, with a budget of $4 million a year to spend on new drugs and therapies for malaria. It is not as much as he would like, but it is better than nothing, and it puts him in the front line of malaria research. Fansidar, halofantrine, mefloquine—you name it, WRAIR has had a hand in nearly every new antimalaria drug to have emerged from Western laboratories in the last thirty years. But so rapid is the spread of resistance that it is a constant battle to stay ahead. Indeed, some experts believe that drug resistance is so prevalent that the world is facing its severest crisis since the discovery of quinine.

On the wall above his desk, Milhous has a flowchart tracing the progress of drugs currently in the pipeline. Malarone—a new GlaxoSmithKline prophylactic against falciparum that combines proguanil and atovaquone—has already been licensed for use in Kenya and is in the final trial stage in the United States. Just behind it is tafenoquine, a new "super prophylactic" that Walter Reed has also developed jointly with Glaxo. Then there are the derivatives of a "new" plant compound, artemisinin: artesunate, artemotil, and a formulation Milhous believes is particularly promising, artelinic acid.

Milhous first heard about artemisinin in 1982 when he was a fellow at Burroughs Wellcome and received a report from the WHO malaria steering group in Geneva about a promising new Chinese plant called *qing-hao*. Rumors of a miraculous new Chinese cure for malaria had been circulating for years. The Chinese had first published their findings in the *Chinese Medical Journal* in 1979, but when the WHO approached Chinese scientists for samples of the plant so that they could conduct their own assay, they were rebuffed. "This was just after the Nixon era, remember. Mao Tse-tung was still in power and the Chinese were very skeptical about sharing information and news," says Milhous.

It turned out that *qing-hao* had been known to Chinese herbalists for more than two thousand years. The first Western botanist to describe it was Linnaeus, who gave it the name *Artemisia annua*, but to the green-fingered it is better known by its colloquial name, sweet wormwood. Like cinchona and other members of the Compositae family, to which it is related (sagebrush, tarragon, absinthe), sweet wormwood is noted for its aromatic bitterness. English herbalists had used it for years, but it was only when, at the height of the Cultural Revolution, Mao ordered his scientists to "solve" the problem of malaria, which was then sweeping China's southern provinces, that *qing-hao* was "rediscovered." It had first appeared in a Chinese recipe book in 168 B.C. as a treatment for hemorrhoids. Then in *The Handbook for Emergency Treatments*, a book written by Ge Hang in A.D. 340, it was recommended as a treatment for chills and fevers. Ge Hang instructed fever sufferers to soak *qing-hao* in one *sheng* (about one quart) of tea, squeeze out the juice, and drink the remaining liquid. Following Ge Hang's recipe, Chinese scientists prepared a decoction and fed it to mice infected with *P. berghei*, a lethal rodent malaria parasite. They found that *qing-hao* was as good as choloroquine and quinine at clearing the parasite. It also cured chloroquine-resistant strains of the rodent malaria. By 1972, Chinese pharmacists had succeeded in crystallizing the active principle—*qinghaosu*, or artemisinin—and Mao's scientists began testing it on humans. Seven years later they reported their findings. According to the Chinese, artemisinin cured falciparum malaria more rapidly and with less toxicity than either chloroquine or quinine. Moreover, it also worked against strains that were chloroquine-resistant.

Artemisinin came as a bombshell to the U.S. military. Just as the Galenists and English Puritans had been reluctant to accept the "popish remedy" three hundred years before, so the heirs to Walter Reed cast doubts on the Chinese claims about artemisinin. How was it possible that a bunch of "reds" had beaten Walter Reed's finest brains to such a significant discovery? The Chinese, they pointed out, had made similar claims in the past—most notably when they said they could cure malaria with acupuncture—only to be proved wrong. Why should they believe the Chinese now, especially when the studies they were publishing appeared to date back nearly ten years?

It was Dan Klayman, then head of medicinal chemistry in Walter Reed's experimental therapeutics division, who decided to take a closer look at the Chinese study, and as he did, his interest grew. Artemisinin was an endoperoxide, a molecule consisting of two oxygen atoms, but in a form he had never seen before. On exposure to air, endoperoxides normally become unstable and fall apart, yet the Chinese were claiming that artemisinin could be crystallized into a drug that would persist in the body long enough to destroy the parasite.

China's refusal to share her techniques with the WHO was understandable given the presence of Walter Reed representatives on the WHO chemotherapy and malaria steering group. Rightly or wrongly, the Chinese feared that if they handed over the formula for artemisinin, Walter Reed would patent it for the benefit of capitalist pharmaceutical firms. Under the circumstances, Klayman had no choice but to gather his own samples of the plant. Thus a new hunt was begun, not for a tree this time but for a weed.

Klayman recruited botanists from the Smithsonian Institution in Washington and began scouring North America for *Artemisia annua*. They found it growing right under their noses at Harpers Ferry, just down the road from Walter Reed headquarters. Klayman had Boy Scouts harvest the plant and then set about duplicating the Chinese extraction process. It took him two years to solve the problem, but in 1984 he cracked it, grabbing the May cover of *Science* with his announcement that artemisinin was a poorly water soluble crystalline compound. The two oxygen atoms, he revealed, were like a bomb waiting to explode. All it took to set them off was iron, and where was iron found? Why, in the malaria pigment of course—the heme the parasite releases when it devours red blood cells. When artemisinin came into contact with the iron-rich heme the molecule fell apart, triggering the release of free radicals that were toxic to the parasite. Indeed, artemisinin was so explosive that Klayman speculated the parasite would have little time to recognize its structure and develop resistance.

Klayman's faith in artemisinin has since been justified many times over. Today artemisinin and other artemether-group drugs are the main line of defense against drug-resistant malaria all along the Thailand–Myanmar border. Unlike quinine, which can be toxic in the doses needed to clear severe falciparum, the artemether drugs are faster acting and have fewer side effects. The most rapid of all is artesunate, a water-soluble derivative of artemisinin originally isolated by the Chinese. Indeed, doctors at the Mae La refugee camp in northwest Thailand—the center of some of the most potent drug-resistant malarias in Southeast Asia—have found that artesunate clears falciparum parasites from the blood faster than any drug they have encountered. Artesunate is manufactured in China and Vietnam, but ironically, only one Western pharmaceutical company has so far been willing to make it. The reasons are all too predictable. "Because the Chinese isolated it first, artesunate is not patentable," says Milhous. "And without a patent, pharmaceutical firms aren't willing to pick up the codevelopment costs." Instead, Walter Reed has concentrated on two other derivatives: artemotil and artelinic acid. The Dutch company ARTECEF registered artemotil in Holland in March 2000, and Walter Reed is hoping that GlaxoSmithKline will soon seek a similar registration for artelinic acid in the United States. In addition, the WHO is seeking licenses for its own artesunate suppository. The significance of the artemether-group drugs cannot be overstated. Aside from intravenous quinine, artemotil is the only treatment for severe falciparum malaria currently licensed for use in Europe. The pity is that it has taken so long—nearly twenty years—to become available.

ACCORDING TO MILHOUS, there are two key lessons to be learned from the postwar history of malaria. The first is that "pharmaceutical companies simply aren't interested in developing drugs for people who cannot afford shoes." In other words, there are so few profits in malaria that in order to have any hope of coming up with therapies for the drug-resistant strains of the future, governments must take the lead in drug development. This is now beginning to happen. In 1999, WRAIR received a "presidential uplift" of $3 million on top of its current budget in recognition of the importance of its work on malaria, and extra money from the National Institutes of Health for the development of tafenoquine. Meanwhile, the WHO, along with the World Bank, the International Federation of Pharmaceutical Manufacturers, the U.K. Department for International Development, Holland, and the Rockefeller Foundation, has established a Medicines for Malaria Venture, with the aim of raising at least $15 million a year for new drug development in partnership with drug companies.

The second lesson is that cutting corners by the money-saving trick of taking a partial rather than a full course of drug treatment is simply no longer an option. Nor does it make sense to mortgage the farm on one drug—as occurred with chloroquine in the 1950s—because overuse is just as likely to lead to the emergence of drug-resistant strains as underdosage. Indeed, most scientists now accept that the therapeutic life of whatever drug they develop is bound to be limited. That is why Walter Reed and others are increasingly advocating drug combinations. Thus, in 1994, when Professor Nick White of Bangkok's Mahidol University and his colleague François Nosten saw that mefloquine was no longer effective in half the cases of children infected with falciparum on the Thailand–Myanmar border, they combined it with artemisinin and then artesunate. Their hope is that by means of the combination of drugs, malaria transmission will be greatly reduced and mefloquine's "useful therapeutic life"—to borrow a phrase coined by White—will be greatly extended. But just in case he is wrong, White is now also conducting trials with artesunate in combination with Malarone, the new prophylactic developed by GlaxoSmithKline, in the hope that because Malarone is so new to the market, there is still time to extend its therapeutic life before it encounters drug resistance.[8] In theory, drug combinations could slow the development of resistance in areas of high malaria transmission by tenfold or more. Says White, "That means you need a new drug only every fifty years, not every ten." The only problem is persuading hard-pressed African governments to accept the additional cost of funding two drugs where one would normally be sufficient to return a patient to health.

Where does all this leave the military? The answer is, not much further advanced than at the beginning of World War II. Then the United States and Britain had only quinine. Now they can deploy a whole range of antimalarial drugs without fear of supplies suddenly falling into enemy hands. But the development of these drugs has not been driven so much by strategic necessity as by the evolution of the parasite. Moreover, no matter how effective the prophylactic, there is still no guarantee that soldiers will take it. As yet, there is no resistance to mefloquine in Sierra Leone, for instance. But when Britain recently sent paratroopers to fight antigovernment rebels there, sixteen fell ill with malaria. The reason appears to be that the paratroopers discontinued the mefloquine rather than endure such unpleasant side effects as dizziness, sleeplessness, and depression. Indeed, one soldier was reportedly driven to such distraction by nightmares and hallucinations associated with the prophylactic that he shot himself. Such problems are by no means confined to the British army. When the United States deployed troops to Somalia during Operation Restore Hope, malaria was the number one cause of casualties. The U.S. Air Force forbids pilots to take mefloquine because of the associated side effects and issues them instead daily doses of doxycycline. But as the United States discovered during Operation Restore Hope, a daily prophylaxis regime is difficult to enforce. Even under the most heavily enforced medical discipline in Somalia, for instance, one marine unit suffered a staggering 10 percent malaria attack rate in a single month.

There is one thing, however, that would guarantee that troops entering a war zone were fully protected against malaria, and that is a vaccine. But the problem is that malaria is a protozoan, not a virus, and so far no one has succeeded in developing a vaccine against any parasitical infection.

12

The Vaccine Hunters

> There's no place on earth that's off the malaria map: Arctic circle, freezing mountaintop, burning desert, you name it, malaria's been there. We're not talking millions of cases here; more like hundreds of millions . . . And even when it's properly diagnosed it's not like quinine is always going to get you home safe. With certain kinds of malaria you can mainline quinine all the livelong day and come nightfall you'll still be gathering freezer-burn in the mortuary.
>
> —AMITAV GHOSH,
> *The Calcutta Chromosome*

IT IS EARLY EVENING in Cartagena and the fifteenth International Congress on Tropical Medicine and Malaria is about to get under way. Inside the modern, air-conditioned convention center overlooking the harbor where in 1741 Admiral Vernon's fleet was repelled by the mosquito, nearly four hundred scientists from countries as diverse as Britain, the United States, Thailand, and The Gambia have assembled to hear Colombian president Andreas Pastrana's opening address. It is the first tropical medicine congress to be convened in South America for many years, and the fact that it is being held in a country racked by violence and political kidnappings is a testament to Pastrana's powers of persuasion. In ten days' time Bill Clinton will arrive in the beautiful and, as the conference brochure puts it, "safe" city of Cartagena to unveil a $1.3 billion aid package aimed at eradicating Colombia's trade in cocaine. But for the moment the focus is on another global scourge—malaria—and on why, despite the availability of more antimalaria drugs than ever before, infant mortality from the disease in Africa has remained virtually unchanged in fifty years.

Pastrana runs through the familiar litany: 300–500 million cases a year, and between 1.5 and 2.7 million deaths, half of them children in sub-Saharan Africa. "Every thirty seconds, somewhere in the world a child dies of malaria," he says, before concluding his speech with a call for the establishment of a global fund to fight the disease.

The delegates applaud politely. Even though there are already many such global initiatives and they have heard the statistics many times before, that does not make the numbers any less shocking. In the last fifteen years, for instance, AIDS has killed about 18 million people worldwide, whereas malaria has killed at least 23 million. Some parasitologists believe that malaria may have killed one out of every two human beings who ever lived. The only diseases that are bigger killers are dysentery and tuberculosis. But, of course, there is a vaccine for tuberculosis, albeit one that is increasingly ineffective in many parts of the tropics. As yet, however, there is no vaccine for malaria and thus little possibility of eradicating the disease. Nevertheless, a number of scientists believe they are very close. Indeed, one man claims he is on the point of cracking the code governing the parasite's ability to evade the body's immune defenses and is just waiting for the final pages of the codebook. His name is Manuel Elkin Patarroyo and he is Colombian.

Patarroyo is making his way through the delegates emerging from the conference chamber now. A short man with unruly gray hair offset by a shiny bald spot, he is barely visible through the crowd thronging the steps. But you can tell it is him simply by the flashbulbs and the number of would-be Colombian beauty contestants trailing in his wake. Imagine Stephen Hawking, the wheelchair-bound English physicist, regularly featured in the pages of *Hello!* or *Vogue* and you get some sense of the media frenzy that attends Patarroyo's every public appearance in Colombia. Dressed in an immaculate blue suit and red tie, he looks more like a banker than a scientist. According to national opinion polls, Patarroyo is the second most prominent Colombian this century, just behind his good friend, the celebrated novelist Gabriel García Márquez. Were he to run for the presidency tomorrow, he would probably be elected.

What makes Patarroyo's celebrity so remarkable is that malaria is not usually considered a glamorous endeavor. Most of the breakthroughs in malaria came at the end of the nineteenth century, as a result of Ross's painstaking dissection of mosquitoes, and Italian researchers' close observation of blood slides. It was another hundred years before the revolution in molecular biology and biotechnology transformed our understanding of the malaria parasite and other infective organisms, and gave researchers new tools with which to probe its defenses. But if you are seeking fame and publicity, far better to involve yourself in a cause that preoccupies the developed world, like the cure for cancer or AIDS.

That Patarroyo has generated similar excitement with his search for a vaccine for malaria is a tribute to both his charisma and the international scientific community's doubts about the rigor of research carried out at his institute in Bogotá. Located in a run-down district to the east of the city center, Patarroyo's headquarters is reached via a path overgrown with weeds that winds to the back of the grimy San Juan de Dios public hospital. One of the three buildings is little more than a shell with pigeons roosting in the eaves; on the others, the once grand porticoes are cracked and the plaster is peeling off. The moment you cross the threshold, however, it is a different story. Laboratory technicians deep in concentration scurry along the corridors, their pockets bulging with computer printouts. On one floor young women are carefully measuring white powders into a series of vials, while downstairs young men jot chemical equations on a blackboard. Meanwhile, in another part of the complex a nuclear magnetic resonance machine is slowly scanning a key molecule of the *Plasmodium falciparum* parasite. In the three-dimensional image being generated on a nearby computer, it looks like something out of a science-fiction film, a space-age insect with color-coded legs.

Sitting in an office at the top of a marble staircase lined with contemporary Latin American art is the reason for all this frenetic activity: Patarroyo, or, as his colleagues prefer to call him, El Jefe—"the Chief." For the past twenty-five years Patarroyo has devoted himself single-mindedly to the search for the cure for malaria. It is the Holy Grail of malaria research—the modern equivalent of the nineteenth-century hunt for cinchona—a one-shot vaccine that will provide complete protection against falciparum, and the other human malaria parasites, for life. From Washington to Oxford to Paris to Sydney, many other scientists are chasing the same goal. Whoever succeeds will almost certainly win the Nobel Prize. But according to Patarroyo, his vaccine is the most advanced. "I am already 80 percent of the way there," he says, handing me a sheaf of papers recording the trials of his synthetic chemical vaccine, SPf66. "All I am waiting for are the remaining rules."

Other scientists beg to differ. In Washington, for instance, the U.S. Army and Navy each has its own advanced vaccine development program, making what Patarroyo claims is a one-horse race at least a three-horse race. And while Patarroyo's vaccine has reputedly performed well in South America, the results of independent trials in Africa and Southeast Asia have been disappointing. But such is Patarroyo's celebrity in Colombia that few of his countrymen doubt him. Just as José Celestino Mutis put Colombia on the map with his botanical institute devoted to the taxonomy of cinchona, so Manuel Elkin Patarroyo's Institute of Immunology has brought Colombia to the attention of the international scientific community. The hundred and fifty scientists working out of his state-of-the-art laboratories have the support of both the Rockefeller Foundation and the World Health Organization. As Howard Engers, the director of the WHO vaccination development program, puts it: "I am very impressed by his work. He is a very motivated, hardworking person, and if he says he has succeeded, we take him seriously."

TO UNDERSTAND WHAT IT WOULD TAKE for a vaccine against malaria to work, you first have to understand how vaccines protect against bacteria and viruses. A virus's sole raison d'être is to colonize living cells. To prevent this, vaccine makers take attenuated or dead versions of the virus and use them to stimulate the immune system to produce antibodies. The same principle is also applied to bacteria. The idea is that if you can train the immune system to produce antibodies against an apparent threat—an attenuated version of a bacillus or a virus—then when the real thing comes along the immune response will be vastly accelerated and therefore able to take the invader out before it has a chance to do any damage. The first and best example is the smallpox vaccine.

In 1796, Edward Jenner, an English country doctor, used cowpox—a disease related to smallpox that causes serious illness in cattle but not in humans—to inoculate workhouse children against its deadly human cousin. Eventually, smallpox was eradicated worldwide. Jenner, however, had no concept of the viral nature of smallpox and only hit on inoculating with cowpox after seeing how milkmaids who were exposed to the bovine disease were protected. The real birth of immunology came with Louis Pasteur's observation in 1887 that you could prevent chickens from acquiring fowl cholera by inoculating them with a dying culture of the fowl cholera bacillus. By the 1960s, Pasteur's insight had resulted in vaccines against a host of life-threatening bacterial and viral agents—from mumps to polio to yellow fever to measles.

Unfortunately, designing a vaccine against malaria is not as straightforward as designing a vaccine against a virus or a bacillus. A virus is simply a small strand of nucleic acid, either RNA or DNA, wrapped in a protein coat. Strictly speaking, it is not alive but co-opts the machinery of living cells to produce copies of itself. A bacillus is more like the malaria parasite, in the sense that both are single-celled organisms. However, while a bacillus is always a bacillus (and a virus always a virus), the malaria protozoan comes in a myriad of guises. It enters your bloodstream as a sporozoite, bursts from the liver as an asexual merozoite, then transforms itself into a gametocyte and jumps back into the mosquito to begin its life cycle again. This constant shape shifting makes it extremely difficult for the immune system to recognize and mount a sustained attack. Normally, antibodies would identify alien invaders in the bloodstream, while killer T cells would hunt down and destroy unwelcome cellular guests. But the sporozoites are only in the blood for half an hour—too short a time for our antibodies to destroy all of them—and by the time the T cells are ready to remove them from the liver, they have transformed themselves into merozoites and migrated to red blood cells, where they are once again safe from attack.

Parasitologists believe that this extraordinary capacity to outwit and evade our immune system is proof of malaria's antiquity and its long association with man. Some biologists believe that falciparum, and the other human malaria protozoa, evolved from a unicellular plant organism, like algae, as its DNA shows that a distant ancestor could manufacture chlorophyll, the substance that makes plants green and allows for photosynthesis. One theory is that the protozoan began as a parasite of a prehistoric plant-eating reptile and was later picked up by a mosquito; another is that it was a parasite of the mosquito first and subsequently became established in dinosaurs. We know from the fact that anophelines transmit more than a 170 similar protozoa to a wide range of vertebrates, from reptiles and birds to rodents and apes, that the parasite has been around for at least thirty million years. More important, from our point of view, we also know that many of the surviving bird and monkey malarias are closely related to the plasmodia that infect us. *P. cynomolgi*, a malaria parasite carried only by the Southeast Asian rhesus monkey, is very closely related to vivax malaria. And the DNA sequences of *P. reichenowi*, a malaria parasite of chimpanzees, are so close to falciparum's they could almost be a forgery—as if someone had copied the gene strings as one would copy a sonata, hitting an occasional false note. Fortunately, over time, man has evolved various strategies for living with malaria, and it is these strategies that hold out hope that Patarroyo and other vaccine makers may one day succeed in cracking the parasite's defenses.

Take sickle-cell anemia, an excruciating and usually fatal genetically transmitted blood disorder found only in black Africans and blacks of African descent. Victims of sickle-cell disease, who inherit sickle-cell genes from both parents, tend to die young, often before they have had the opportunity to bear children. Natural selection should therefore have led to the disappearance of sickle-cell genes, as those who develop the disease are less likely to reproduce than other members of the population. But this is not what happens. Instead, many African blacks still carry the sickle-cell "trait." The reason is that people who get the gene from just one parent have partial protection against malaria: they can still be infected with falciparum, but their sickle-cell gene prevents them from developing the full-blown version of the disease. In areas where malaria is endemic, therefore, infants with the trait have a greater chance of survival than those without it and thus have a greater chance of producing offspring who also carry the aberrant gene.[1]

Another reason for optimism is that if you survive malaria in infancy and continue to be exposed to it in early adulthood, you will eventually develop partial immunity to the parasite. This is because repeated exposure to the disease keeps the immune system in a high state of alert, so that there are always antibodies ready to attack the parasite as soon as it invades. People with acquired immunity can still be infected with malaria and transmit it, but their symptoms are milder, with the result that the disease is rarely fatal. The only drawback is that the immune response is strain-specific. A Gambian teenager may be immune to the strains of falciparum in his village, but if he travels to the village next door, he may encounter a different strain and fall ill. These evolutionary differences between strains of the same parasite may represent the most formidable challenge facing Patarroyo and other immunologists.

To have any hope of training the body to fight malaria, vaccine makers must first decide which stage of the parasite's life cycle they wish to concentrate on. For example, you could develop a vaccine that stops the plasmodium from infecting liver cells—the sporozoite stage—but has no effect on merozoites already in the blood. Or you could develop an asexual vaccine for the blood-stage parasites that would have no effect on the sexual gametocytes that actually transmit the disease. Finally, you could design a transmission-blocking vaccine that stimulates antibodies against gametocytes and stops mosquitoes from being infected but has no effect on either sporozoites or merozoites. Moreover, whichever stage you decide to concentrate on, the vaccine has to be perfect. It's no use making a vaccine that kills most of the sporozoites; it takes only one parasite to reach the liver and begin manufacturing merozoites for your "sporozoite specific" vaccine to be useless.

For various reasons, Patarroyo and other researchers decided to concentrate on the sporozoite stage. First, it was the most direct approach: a vaccine that stopped the sporozoites from reaching the liver would kill infections at the source and thus provide absolute protection both for nonimmune travelers to malaria-endemic areas and for children most at risk of dying from the disease. Second, and more important, by the early 1970s scientists knew through experiments with radiation that it was possible to use attenuated sporozoites to confer complete protection.

In 1968 researchers at New York University found that if mosquitoes infected with malaria were dosed with X rays, the sporozoites in their salivary glands were inactivated: that is, they were still alive and capable of eliciting an immune response, but they were no longer able to invade host cells. Having ascertained that the irradiated sporozoites conferred immunity in rodents and monkeys against their malarias, U.S. Army scientists began a series of experiments in 1971 to see if they would work in people too. It took several trials, but eventually the Walter Reed institute confirmed that if volunteers were bitten at least a thousand times by irradiated mosquitoes, they were protected against malaria for up to six months, after which the immunity wore off. Unfortunately, the protection was strain-specific. Moreover, if you extracted the irradiated sporozoites from the mosquitoes, the sporozoites died, and when you injected them, nothing happened. In other words, you had to use live mosquitoes. Building a massive insectary to deliver the vaccine was obviously impractical. However, by now researchers had noticed that the key to the immune response was a series of proteins on the skin of the sporozoite. One, the circumsporozoite (CS) protein, seemed to be particularly promising. If it could be cultured biologically, or, better still, synthesized chemically, you might be able to skip the mosquito altogether and manufacture sufficient quantities to make a vaccine viable.

By the early 1980s, biotechnology had progressed to the point where you could extract any gene—whether from brewer's yeast or a nematode worm—and engineer a bacterial culture to "express" it. In theory, this meant you could grow the CS protein in a culture and then insert it into an inert virus—say, smallpox or hepatitis—and see whether you got an immune response. Similarly, the new peptide-synthesizing machines meant that if you knew the sequence that formed the protein along any section of the parasite's skin, you could manufacture the same protein molecule chemically. This is exactly what Patarroyo and other researchers decided to do when it was discovered that the key CS antigen consisted of a sequence of just four amino acids repeated over and over again.

IT IS NOT SO MUCH Patarroyo's fame that has brought me to Bogotá as the intriguing parallels between him and José Celestino Mutis. Although Mutis is better known as a botanist and an early Colombian champion of Newtonian and Copernican physics, he was first and foremost a physician. Like Patarroyo, he was fascinated with disease. As well as experimenting with quinine to cure malaria, Mutis conducted the first South American trials on smallpox, inoculating patients in Cartagena with pus taken from the sores of smallpox victims in an attempt to provoke an immune response. More important, like Patarroyo, Mutis put Colombian science on the map. At his herbarium, which housed twenty thousand different New World plants, he oversaw a team of thirty draftsmen. Like the chemists and immunologists at Patarroyo's institute, they worked in shifts, sketching in painstaking and exquisite detail the different species of cinchona growing in the *montaña* of Colombia and Ecuador. It was Mutis's descriptions of cinchona that enabled Linnaeus to complete his classification and led him to describe Mutis as "*nomen immortale.*" Had it not been for Mutis, it would have taken very much longer for the world to discover the valuable new Colombian species of cinchona. And had it not been for Mutis, it is unlikely that Humboldt—the most eminent scientist of his day—would have felt compelled to visit Bogotá and linger as long as he did.

Today, it is not uncommon for eminent scientists and elder statesmen to make a similar pilgrimage to the San Juan Hospital. The walls of Patarroyo's office are cluttered with pictures of honored guests: Nelson Mandela, the king and queen of Spain, the president of Nigeria, and my own favorite, Princess Anne. Then there are the awards: the Robert Koch medal, the Edinburgh medal, the French Legion of Honor . . . the list goes on and on. As we sit talking, the fax machine vibrates into life, and out pops a message from the mayor of a town in southern Spain. "They want to name a street and their new biotechnology institute after me," says Patarroyo proudly. "I must get two or three proposals like this a week."

The institute is a world apart from Ataco, the small town in the Tolima region of Colombia where Patarroyo grew up. His parents were middle-class businesspeople who expected their eleven children to follow in their footsteps. Instead, they ended up with five physicians, one nurse, and a child psychologist, among other professionals. Patarroyo was the eldest, and in his case, the catalyst for his career was an illustrated science book his parents gave him when he was eight. "It was all about the discoveries of Pasteur, Koch, and Ross," he recalls. "I was captivated. Some people want to be football stars. I decided then and there I wanted to be an immunologist."

Patarroyo attended medical school in Bogotá, but by his own admission, he was a mediocre student. Then, in the second year of his studies, during his internship at the San Juan Hospital, he met Ronald Mackenzie, the head of the Rockefeller Institute, and told him he was interested in pursuing a chemical approach to vaccines. In 1965 this was a radical idea. Ever since Pasteur, all vaccines had been cultured biologically. But what, wondered Patarroyo, if you could achieve the same effect by synthesizing key immunological antigens chemically? Then you would no longer have to worry about taking living material and attenuating it. Moreover, the resulting vaccine would be a lot cheaper to produce.

With this end in mind, Patarroyo went to Yale University in 1968 on a scholarship to study virology. Next he joined the immunology department of Rockefeller University. It was there that he came under the sway of Bruce Merrifield, the chemist whose pioneering work on synthetic peptides would lead to a Nobel Prize in 1984. "Merrifield was the single biggest influence," says Patarroyo. "He was the one who convinced me that the synthetic approach could work." However, it was only when Patarroyo returned to Bogotá that he decided the vaccine had to be malaria.

Patarroyo switches on a slide projector beside his desk. "Look, the biggest killer worldwide is tuberculosis," he says, pointing to a pie diagram divided into disease "slices." "Next is dysentery. But the third biggest is malaria." He clicks to the next slide, of an African baby with weeping eyes. "Every year between one and a half and three million people, one third of them children in sub-Saharan Africa, die. The reason is that the principal drug, chloroquine, no longer works, and resistance is growing to the alternatives. A vaccine is the only hope."

In the mid-1970s Patarroyo persuaded the San Juan Hospital to give him his own immunology wing, funded by money from his private practice, and he began exploring ways to target the parasite. At this stage he was by no means the leader in the field. In 1979 the husband-and-wife team of Ruth and Victor Nussenzweig—the researchers at New York University who had conducted the first experiments with irradiated sporozoites in mice and monkeys—identified the CS protein as the most likely target of a vaccine. A few years later they succeeded in synthesizing the protein in the lab and injecting it into mice. Not only did the mice produce antibodies against malaria, not a single sporozoite survived to colonize the rodents' liver cells. Excitement ran high. It looked as though the Nussenzweigs had solved the problem. However, the human trials in 1987 turned out to be a huge disappointment. Only one of three volunteers immunized by the Nussenzweigs remained free of malaria. In the case of the other two, the vaccine delayed the onset of the disease but failed to stop the emergence of merozoites. By now, Walter Reed and the National Institutes of Health had also entered the stakes. Teaming up with one of the world's biggest pharmaceutical firms, SmithKline Beecham, they tried a different approach. Instead of synthesizing the CS protein, they extracted the gene that encoded the protein from the parasite, inserted it into bacteria, and let the bacteria generate copies of it. They then injected the CS protein into six volunteers. Again, only one remained disease-free. The other five developed malaria, albeit more slowly than they would have without the vaccine.

The failure set back the cause of malaria vaccine research in the United States for years. The problem was that expectations had been pitched unrealistically high. In truth, the results were encouraging, but because the failure had been so public, drug companies retreated to lick their wounds, and the sources of research funding dried up. Patarroyo, however, was not hampered by such constraints. Just as

Mutis answered only to the viceroy of New Granada, so Patarroyo answers only to the president of Colombia. As he told me, "If I want something, I simply call up Pastrana. Not even the minister of health can tell me what to do."

In the eyes of the international scientific community, this strength is also Patarroyo's weakness. Unlike his rivals at more staid institutions in Washington and Oxford, Patarroyo is not subject to continual peer review. He can pursue a line of inquiry—possibly the wrong line of inquiry—for years without having to publish his findings and run the risk of being challenged. To his supporters, this makes him like the gentlemen scientists of the nineteenth century. After all, would Darwin have arrived at his theory of natural selection if he had had to go cap in hand to the Medical Research Council for a travel grant to South America? But to his detractors, Patarroyo's methods are risky and wasteful. To them, his empirical approach to the vaccine—altering an amino acid here, an amino acid there—seems to have more in common with alchemy than science.

Nevertheless, in the early 1980s Patarroyo began churning out hundreds and then thousands of different amino acid sequences, turning down the offer of an assistant professorship at the Rockefeller Institute in New York in order to concentrate on his own research in Bogotá. In 1987, at around the same time that malaria vaccine trials were getting underway in the United States, he reported that he had come up with his own synthetic vaccine, SPf66, that conferred 50 percent protection in monkeys infected with falciparum. Then, in the first human trial, involving fifteen Colombian army volunteers, he reported that three out of five who received the vaccine were protected. Between 1988 and 1993 a series of major trials were conducted in Colombia, Venezuela, Ecuador, and Brazil. The results were confusing. In some, the vaccine seemed to have an efficacy rate as high as 80 percent; in others it was as low as 35 percent, or else the results were statistically insignificant. Nevertheless, turning down lucrative offers from drug companies, Patarroyo donated the rights to SPf66 to the World Health Organization in 1995, boasting that the problem of malaria had been "solved."

Once again it was a false dawn. In randomized, double-blind trials sponsored by the WHO in The Gambia and Tanzania, SPf66 showed no protection in children under the age of five. Then, in a trial sponsored by the U.S. Army in Thailand in 1996, twelve hundred Thai children given several inoculations of SPf66 were no more protected than those given a placebo. The verdict of the international scientific community was unanimous: SPf66 did not work.

To any other scientist such a setback would have been devastating. Patarroyo was roasted for his methodology and for the ethics of moving so quickly to human trials. His critics suggested that the reason he had obtained such promising results with his Colombian army volunteers was that he had "challenged" the vaccine by injecting his study group with falciparum rather than by allowing them to be bitten with malaria-carrying mosquitoes. Moreover, critics said that many of the Latin American trials had not been properly randomized, that they had inappropriate controls, and that neither his subjects nor the observers were blinded. Patarroyo quickly acknowledged that he had "made mistakes" and had much to learn about epidemiology, but he was adamant that he was on the right track. Hitting out at the "intellectual racism" of scientists in the North, Patarroyo argued that challenging the vaccine by intravenous injection was more accurate because it was impossible to tell in what quantities the parasite was harbored by mosquitoes. Believing that the vaccine's poor performance in Africa resulted from his failure to take into account variations in local African strains of falciparum, he retreated to his peptide synthesizer in Bogotá, determined to do better. Three years later, in 1999, he emerged from his self-imposed purdah to tell a British newspaper that he had developed an improved version of SPf66 and was ready to conduct new trials. Within two years, he confidently predicted, his vaccine would be available "worldwide."

Sitting in his office in Bogotá a year later, Patarroyo remains acutely sensitive to criticism. He is still smarting from the conference in Cartagena, where he claims members of the British media had been warned by other delegates not to take his claims about the vaccine "seriously." "I don't mind being questioned by my peers. That's natural, I'm a scientist. But when people come at me with an agenda and don't tell me, that's unfair." As he says this I cannot help noticing a sketch on the wall depicting Patarroyo felling the malaria "beast" with a hypodermic. The drawing bears the slogan "*David y Goliat*" and the inscription "*Para el doctor Patarroyo con mucho humor.*" Humorous or not, don't such impressions play into the hands of his enemies?

Patarroyo admits that designing a vaccine has been very much more complicated than he, or anyone else, had imagined. "But when we say there's a malaria vaccine, we are not saying, Bingo! here is something we have just taken off the shelf. It has taken a hell of a lot of work to get this far."

Patarroyo argues that the fundamental reason for SPf66's disappointing performance in Tanzania, The Gambia, and Thailand is the variations in the African and Southeast Asian strains of falciparum. These variations, he believes, make all the difference between a vaccine that works in the laboratory and one that performs in the field. His solution is to target sections of the key binding proteins on the surface of the blood parasite that are "conserved by" (common to) different genetic strains of falciparum. Unfortunately, these conserved sequences are also the ones associated with the weakest human immune response. To get around this problem, Patarroyo has been inserting strongly immunogenic peptides at different points in the proteins in the hope of engineering new molecules that are immunogenic against *all* strains of falciparum.

"Imagine the parasite like a human hand and the conserved sequences like fingers," suggests Patarroyo, reaching for the door to his office. "The parasite wants to unlock the door and find a way into the host cells. Our aim is to stop it, to keep the door locked. The problem is, the parasite can open the door with these fingers"—Patarroyo grabs the door with his right hand—"or with these fingers"—he grabs the handle with his left hand instead. "My job is to train the immune system to recognize which hand the parasite is using so that its fingers no longer work."

Since 1995 Patarroyo has synthesized thousands of peptides in search of the right mix of conserved and immunogenic sequences. He really believes that the conserved regions are governed by certain "rules" that he is on the point of divining. If he can find just the right cocktail of antigens, Patarroyo believes, he will have the equivalent of molecular gold: a vaccine that confers 100 percent protection against any strain of the parasite anywhere in the world. At least, that is the theory.

IT IS 5:30 A.M. at Bogotá's El Dorado airport, and Patarroyo is nowhere to be seen. He has told me to meet him bright and early at the Aero República ticket desk so that we can be sure of boarding a flight to Leticia in the Colombian Amazon. But the only other people in the concourse are a family of Amazonian Indians and an old man laden with boxes packed with rope, fabric, and canned food—essential jungle supplies.

Leticia is where Patarroyo keeps the owl monkeys that are critical to his experiments with the vaccine. The monkeys are indigenous only to South America, from Panama to northern Argentina. More important, the owl monkey is one of only three primates that can be infected with falciparum. According to Patarroyo, the owl monkey's immune system is nearly identical to our own, making it the "perfect" animal model for testing his vaccine. If SPf66 clears falciparum from the blood of owl monkeys, then it will clear the parasite in humans. Like Dr. Moreau, Patarroyo even has his own island—Isla de Los Micos—dedicated to proving his thesis. The island is on the Amazon at the border of Colombia and Peru. But to get there, first you have to fly to Leticia, an Amazonian trading town farther downriver.

It has taken months of badgering by phone and fax to persuade Patarroyo to make the trip. First he was preoccupied with the congress in Cartagena, then he had to attend a banquet in honor of Clinton, and his daughter's wedding will take place in a week's time. The only reason he has agreed to fly with me today is that he has a new batch of vaccine to deliver and is anxious to get the results. But as the airport comes to life and the line at the Aero República desk lengthens, I am beginning to think he has had second thoughts. The flight to Leticia leaves at 7 A.M. and it is now 6:30. I am about to relinquish my place in line when I spot a man, with a baseball cap pulled tightly around a shock of frizzy gray hair, scoot across the concourse and duck behind the ticket desk. The man looks like Patarroyo, but I cannot be sure. It is only when he removes his cap and the attendant, in a sudden flash of recognition, jumps to attention that I recognize him, too. "We were up until ten-thirty last night analyzing new ideas," says Patarroyo apologetically. "The team wanted to know what the boss was cooking up."

The young woman at the ticket desk hands us our boarding passes ahead of the old man with the boxes, and we begin making our way toward the departure lounge. As we wander through the concourse, I begin to understand Patarroyo's earlier furtiveness. On the way, we are stopped by boys wanting his autograph and by mothers wanting him to bless their babies. He deals with each supplicant good-naturedly. His fame in Colombia is such that it crosses the political divide, too. A few years ago, he tells me, he was taken hostage by the Fuerza Armadas Revolucionarias de Colombia (FARC)—the Marxist guerrillas who control the demilitarized zone in southern Colombia. But after just five hours they let him go. "They wanted to tell me that malaria was a problem for their people, too. They now send me faxes urging me to keep up the good work."

Wherever Patarroyo goes, his ledger book, containing the results of his vaccine trials, travels with him. Safely on board the plane, he opens it to show me. Each page is divided into neat rows and columns. The rows record the numbers of monkeys in each trial; the columns show the time it has taken for the vaccine to clear the parasites from their blood. The monkeys are given SPf66 in three doses over four weeks. Then, on day thirty, Patarroyo "challenges" their immune systems by injecting them with falciparum and waits to observe the effect on the parasites in their blood.

"See, in 1994 we gave the vaccine to fifty monkeys. But by day five after infection, all showed signs of parasitemia. As soon as they reach 6 percent parasitemia, we stop the experiment, treat their malaria, and release them. After all, we don't want them to die."

Patarroyo turns to the next page, headed "October 1994." "I thought that maybe the problem was that the vaccine wasn't big enough, so I added more peptides to make the molecule bigger, but that didn't work either." He flips the page, running his finger along each row. "See, every monkey comes down with malaria on day five, so I randomly changed the sequence of amino acids." He turns to January 1995 and points to a row of zeros. "Look, for the first time I began to see zero parasitemia in a few monkeys. That was exciting. It meant we were on to something."

All the monkeys with zero parasitemia subsequently went on to develop full-blown malaria, but it was a significant development: the onset of the infection had been delayed. Working on a hunch that the key was the conserved protein regions, Patarroyo continued mixing up the amino-acid sequences in the hope of finding a cocktail of antigens that would trigger a fully protective immunological response.

For two years he admits he was "working in the dark." Then, in 1997, he started to see zero parasitemia up to day ten in a number of monkeys, and in a few cases the monkeys never went on to develop an infection. Patarroyo was now convinced that—as he puts it—he had discovered "half the rules." But he still had a long way to go before he would be in a position to publish, let alone contemplate human trials. Instead, he continued playing with the amino-acid sequences. By 1999, he had synthesized some fourteen thousand peptides. Then, in June, he claims he had the breakthrough he'd been waiting for. "Look, row after row of zeros." Patarroyo flips the page. "In August it's the same, and in September." By autumn 1999, Patarroyo says, only 20 percent of the monkeys were developing infections. "The vaccine was providing 80 percent protection after three shots," he says.

Since then, Patarroyo has refined the vaccine further, and he now claims he is able to achieve the same results with just two shots. However, he will only be satisfied when he has a one-shot vaccine providing 100 percent protection.

Gradually the Cordillera Central gives way to the Cordillera Oriental, and then the mountains recede from view and we are flying over virgin jungle. Here and there the green canopy is broken by a snaking channel or the metallic glint of a lake, but after about an hour a haze envelops the forest and all I can see for miles in any direction are puffy white clouds. In the wet season Leticia is highly malarial, but it is August and Patarroyo has assured me I do not need a prophylactic. What does he take for malaria? "Nothing. I vaccinated myself in 1986 and haven't needed to worry since. It doesn't matter whether I am in Colombia or Tanzania, I never get ill." Indeed, such is Patarroyo's confidence in SPf66 that he has vaccinated his whole family. So far, he insists, there have been no adverse side effects.

Two hours later the plane dips beneath the clouds and we begin the descent to Leticia. Patarroyo leans across the aisle and points out the starboard window at a mist-shrouded island far below. "Isla de Los Micos," he whispers.

After Bogotá, the heat of the Amazon is oppressive. Raul Rodríguez, the manager of Patarroyo's research station, is at the airport to meet us. A short, chubby man with gentle brown eyes, Rodríguez is dressed for the tropics in shorts and a tennis shirt, whereas we are still bundled up against the chill of the airport at seven in the morning. We quickly strip down to our shirtsleeves, and I hop onto the back of Patarroyo's scooter for the five-minute drive into town. "There are only seven miles of road in Leticia," explains Patarroyo over the hum of the bike. "This is the only way to get around."

Patarroyo drops me at the Hotel Anaconda—a huge air-conditioned palace overlooking the town square. The manager greets him warmly. Our arrival has already created quite a stir. As we stroll to a nearby restaurant for breakfast, Patarroyo pauses to shake hands with everyone from the corner cigarette seller to the manager of the local bank. By the time we sit down for *caldas* and hot chocolate, we have picked up a conga line of shoeshine boys and are dripping perspiration. Fortunately, it is only a ten-minute walk to the research station.

On the way, Rodríguez fills me in on *Aotus*. It is on the protected primate list and is the only truly nocturnal monkey in the world. "That's why we call it the owl," says Rodríguez. "We feed them eggs, papayas, bananas. You will see, they are very well cared for."

In the wild, the monkeys sleep high up in trees, nestled in tangles of vines. The only way to catch them is to climb into the canopy and snag them with nets. It is a highly prized skill, and one that is passed down from father to son. "Here it costs me $40 to $50 for each monkey," boasts Patarroyo. "In America the same monkey would cost nearly $3,000."

Patarroyo argues that this cheap and plentiful supply of primates gives him a huge advantage over his rivals. Whereas the U.S. Army and Navy breed all their experimental monkeys in captivity, Patarroyo literally plucks his from the riverside. Moreover, because he argues that the monkey and human immune systems are virtually identical, he believes his results will automatically hold good for humans.

Patarroyo's critics disagree. They point out that his snatch-and-grab approach to animal testing means that he has no idea whether the monkeys in his experiments have already been exposed to malaria in the wild. Therefore, he has no way of knowing if some of the monkeys already have antibodies against falciparum—a factor that could influence the outcome of his trials. Moreover, there is evidence to suggest that some owl monkeys are not susceptible to falciparum infection at all. This means that SPf66 may not be protecting as many monkeys as Patarroyo thinks. Conversely, it is possible that a vaccine that has no protective effect in owl monkeys could be protective in people. But the most damning criticism of Patarroyo's monkey trials is that so far other experimenters have been unable to repeat his results. In 1990, researchers at the Centers for Disease Control in Atlanta inoculated monkeys with both their own version of SPf66 and a batch provided by Patarroyo. Neither of the groups was protected. Then, in 1992, another Colombian researcher inoculated monkeys with the same batch of vaccine Patarroyo had used on the Colombian army volunteers. This time the monkeys did not even produce detectable antibodies.

Patarroyo is unmoved by such criticisms. He points out that owl monkeys live in small, monogamous groups and that he deliberately selects his test subjects from different families so as to ensure as much diversity as possible. Moreover, he claims that his vaccine did not perform well in human trials in Africa and Thailand not because of differences in the monkey and human immune systems, but because of differences in the African and Southeast Asian strains of falciparum. Finally, he claims that the latest batch of SPf66 is more potent than previous versions. "We have manufactured a completely new molecule. One that has never before been seen in nature."

We are now standing in front of a pretty walled compound. As Rodríguez opens the gate, I feel as though I am about to step onto a set from *Planet of the Apes.* A sign above the entrance reads "Experimental Primate Research Station." The grounds are brimming with bougainvillea, palms, and ficus trees. To the left is a conference room; to the right, the laboratories and monkey barracks.

Rodríguez unlocks the door and shows me around. The monkeys are housed in small metal pens arranged one on top of the other in long, neat rows. They have just had breakfast. They are smaller than I expected, with grayish brown coats, pale underbellies, and long, delicate fingers, but their most distinctive feature is their large round eyes with white patches above the lids. The patches create the illusion that they are awake even when they are asleep—an important defense mechanism in the wild, presumably, but useless here.

Rodríguez opens the door to a cage and pulls one out. The creature emits a series of high-pitched screeches, but it quickly calms down once it realizes it cannot escape. Next door, Patarroyo is already unpacking his vials of vaccine. The largest room contains 260 monkeys, the one next to it, 100. But today Patarroyo plans to inject just 50. Rodríguez takes the seat beside him. Breaking the seal on a vaccine bottle, Patarroyo sucks the fluid into a needle and, instructing his assistants to hold the monkeys outstretched on the table in front of him, begins injecting the vaccine into their soft undersides.

It takes an hour to vaccinate the 50 monkeys. As each one is brought through, Patarroyo makes a careful note of its serial number in his ledger. Tomorrow Rodríguez will vaccinate a further 200. Then, in twenty days' time, all 250 monkeys will be given a second dose of the vaccine, ready to be infected with falciparum ten days later. "In forty days I will know the results," says Patarroyo.

IT IS NOON and Patarroyo's work for the day is almost over. He has only one more duty to perform: to return fifty monkeys from the last trial to the wild. While Rodríguez prepares the motorboat for the ride to Monkey Island, Patarroyo shows me round his private suite. His bedroom is simple but elegant: two low wooden beds, a writing desk, and a large cupboard full of freshly pressed white shirts. "This is where I count the blood cells when the results come back—ten thousand per monkey. It takes all day."

Patarroyo changes into shorts and a loose-fitting shirt for the boat trip and tells me to help myself to a pair of deck shoes. Then he selects a baseball cap from the cupboard and we are ready to go.

The government gave Patarroyo the island in the late 1980s. In effect, Isla de Los Micos is a giant segregation compound, a way of ensuring that monkeys with vaccine-enhanced antibody levels do not become mixed up with the immunologically virgin monkeys Patarroyo needs for his trials. The only problem is that the island is overcrowded. Patarroyo has been vaccinating an average of 50 monkeys every three months for thirteen years. That makes 250,000 monkeys since 1987. "We have asked the government for another island, but they are not that easy to come by. We will just have to wait and see."

It is now a perfect Amazonian afternoon and blisteringly hot. The jetty is only ten minutes from the research station, but by the time we reach the boat, we are soaked through. A large gray gunship lies moored in midriver, marking the three-way border between Colombia, Peru, and Brazil. Even in a baseball cap and shorts, however, Patarroyo is instantly recognizable and the customs guards wave us through with minimum fuss.

I notice that Rodríguez has loaded the monkeys into canvas sacks and dumped them on the floor of the boat. The temperature inside must be stifling. The bags bulge and leap about the deck as the monkeys wriggle for air and space. The island is thirty-seven miles upriver. In the water bus the journey takes an hour and a half, but Rodríguez assures me that by launch it is just forty minutes. He cranks the motor and the nose of the boat leaps up into the water. Then another crank and we are off. He is right, the boat is very fast, and the breeze makes the heat less oppressive.

As the jungle passes by, it is hard to believe we are conducting an experiment that may transform the lives of children on the other side of the world. Music from a cabana on the Peruvian side of the river mingles with the steady drone of the motor, lulling us into a trance. If the current batch of vaccine performs as well as Patarroyo expects, he plans a human trial in Colombia later this year, followed by larger field-scale trials in Tanzania in 2001. "We are very close," he tells me, admiring the wake from the boat. "I can feel it." For a moment, I think Patarroyo is talking about the island. Then I realize that he is talking about his vaccine. Despite all the disappointments and criticism, his faith in SPf66 is unshaken.

The next thing I know, we are skirting the island, and Rodríguez is looking for a beach on which to land. Most of the banks are overgrown with giant lianas, but eventually we find a clearing and Rodríguez parks the launch. A macaw sticks its beak out of the jungle, curious to see what the commotion is. The air is thick with mosquitoes and the croaking of tree frogs. We form a chain and gingerly begin passing the sacks up the bank. The wriggling has grown more intense, and it is an effort to keep my balance. As the sacks reach the shore, Rodríguez unties them and shakes free the contents. The monkeys shoot out with startling speed, scattering high into the treetops. Within minutes, all the bags have been emptied, and the job is done. Just one monkey seems reluctant to leave. He is lying on his side, breathing heavily. Rodríguez reaches out to touch him, but the monkey shoves his hand away like a petulant child. He is hurt. "*Son los machos*," murmurs Rodríguez. "The males are very territorial."

We wait for the monkey to catch his breath, then prod him again. Slowly, like a wounded trooper, he gets to his feet and limps into the jungle, casting us accusing glances as he goes. "Poor monkey," I murmur. Patarroyo gives me a cold, clinical look. "Think of the children dying of malaria every thirty seconds in Africa. That's who you should reserve your sympathy for."

The next morning Patarroyo boards the first flight out of Leticia. It will be forty days before the results of the trial are known, and in the meantime, he has appointments to keep in Bogotá. I am eager to question him in more detail, but we have run out of time. Throughout the trip he has been dropping constant hints that he has already found the magical mix of antigens and is just waiting for the right moment to publish. But now that he is about to board the plane, he is suddenly reticent. "It's no use changing the molecule anywhere," he says enigmatically. "You have to know exactly where to make the change, otherwise the vaccine doesn't work. That's the beauty of chemistry—it is so precise."

PATARROYO IS NOT the only scientist who believes he is close to a breakthrough. At the Cartagena conference, Stephen Hoffman, a U.S. Navy clinician and immunologist, also announced he had a new vaccine going to trial. Based at the U.S. Navy Medical Research Center in Bethesda, Maryland—part of the same complex that houses WRAIR—Hoffman comes out of the same tradition of military doctoring that produced Walter Reed and William Gorgas. But whereas Reed and Gorgas fought malaria with petrol and quinine, Hoffman is deploying the most sophisticated weapon in the modern immunologist's arsenal—DNA.

A tropical medicine specialist with a passing resemblance to the actor Danny Kaye, Hoffman joined the navy in 1980 after stints in Bogotá, working with malnourished children, and at the London School of Hygiene and Tropical Medicine—the school where Manson demonstrated the mosquito–man–mosquito transmission of malaria. Hoffman's first posting was the U.S. Navy medical research laboratory in Jakarta. It was supposed to be a three-year association. Twenty years later, he is still wearing the uniform.

It was in Indonesia that Hoffman found himself drawn to the study of malaria. He soon familiarized himself with the history: how the extraction of quinine in the early nineteenth century and the synthesis of chloroquine had led to the retreat of the disease, only for it to rebound spectacularly after World War II. How despite the arsenal of therapeutics at the navy's disposal, the marines still lost more men to diseases including malaria than to enemy fire. And how a viable vaccine could eliminate the problem of prophylaxis indiscipline at a stroke. Then he looked at deaths from malaria in Africa. "I studied the figures and realized that in the past fifty years no drug has been developed that has reduced the mortality rate from severe malaria in hospitals in the developing world one percentage point," Hoffman says. Realizing that a vaccine that protected marines could also have incalculable benefits for African children, he decided to learn as much as he could about biotechnology and molecular immunology.

Hoffman joined the staff of the Navy Medical Research Center but found himself spending most of his time at WRAIR, where the efforts to develop a sporozoite vaccine were just beginning. It was an exciting time. Ripley Ballou, a Walter Reed specialist in infectious diseases, had just teamed up with the NIH and SmithKline to conduct the first human trial of the CS vaccine using the latest gene-cloning technology. In the best tradition of army medicine, Ballou and Hoffman decided that the way to test the vaccine was on themselves. These were the first human trials, which, along with the NYU experiment, were to go down as "failures."

In 1987, Hoffman, Ballou, and four other volunteers took the vaccine and allowed themselves to be bitten repeatedly by infectious mosquitoes. After two weeks—the period it normally takes for the falciparum parasites developing in the liver to invade the blood—three of the volunteers fell ill. But not so Hoffman and Ballou, who believed they were protected. Hoffman flew to San Diego, where he was scheduled to deliver a lecture. Ballou went for a run, after which he attended a party. No sooner had he arrived than he fell into a fever and had to leave. That evening his temperature soared to 104 degrees. There was no doubt about it, he had malaria, and the next morning he checked into the hospital.

Meanwhile, in San Diego, Hoffman was enduring a more public torture. "I was in the middle of my presentation when I developed the malaria rigor and the shaking chills. That was it."

Although, as in the NYU trial, only one volunteer was fully protected against malaria, Hoffman and Ballou took encouragement from the fact that they had developed the disease far more slowly than they would have without the vaccine. "We proved the principle that you could make these vaccines and you could induce protective immunity," says Hoffman.

Undeterred, Hoffman and Ballou decided to see whether the antibody response could be boosted by repeated exposure to malaria in the field. Again the results were disappointing—the vaccine resulted in more antibodies but did not confer protection. Demoralized, Hoffman and Ballou retreated to their animal models. Like Patarroyo, Hoffman conducts tests with owl monkeys. Some are kept at the joint army-navy building in Maryland, others at the navy's jungle research laboratory in Iquitos, in the Peruvian Amazon. The difference is that the monkeys are bred from birth in captivity so that their full medical history is known. However, because of the costs involved, most of the trials are conducted on mice infected with rodent malarias.

By the early 1990s, these animal experiments were suggesting that the production of antibodies was only one of the ways that irradiated sporozoites generated an immune response. Another was by stimulating killer T cells to attack the parasites in the liver. The problem was that the recombinant and synthetic peptide vaccines so far developed were incapable of marshaling those T cells into action. Then Hoffman heard about a new technology, DNA vaccines. Often called gene vaccines, they consist of a gene from the pathogen that is stitched into a circular stretch of bacterial DNA known as a plasmid. In theory, all Hoffman had to do was select DNA from the parasite that encoded for key antigenic liver-stage proteins, splice them into a plasmid, and inject it directly into muscle tissue. The muscle cells would then transcribe the DNA into RNA and begin generating the same proteins the parasite produces when it is in the liver. When the parasite next came along, the immune system would recognize its proteins as invaders and send in the marines—in this case, the killer T cells. "In essence, DNA vaccines harness the machinery of the human cell and use it to stimulate the appropriate immune response for us," explains Hoffman.

In 1995 Hoffman's navy research team began testing a DNA malaria vaccine on animals for the first time. The results were encouraging. When they challenged mice with the most infectious strain of rodent malaria, the mice showed no signs of disease. Next they moved to monkeys, with similarly impressive results. Then, in 1997, they conducted the first safety trials on twenty human volunteers. Not only was the vaccine well tolerated, but the majority of volunteers had heightened T-cell responses. The next stage will be to see whether the vaccine actually confers protection when volunteers are challenged with malaria.

The advantage of DNA vaccines is that they are relatively easy to manufacture and can be tailored to be as complex as the parasite itself. In theory, if enough genes from the parasite are inserted into the vaccine, the immune system will recognize the parasite at every stage of its life cycle, defeating its ability to change shape and evade the body's defenses. So far, Hoffman's team has produced cocktails with genes from five, eight, and fifteen different antigens. As yet, however, no one knows whether they will result in sufficiently large T-cell responses in humans.

Meanwhile, Walter Reed, in partnership with SmithKline, had been concentrating on boosting the antibody response of the CS protein using what Ballou calls molecular biology tricks. In 1997 this effort resulted in RTS,S, the most promising vaccine yet. Like SPf66, the vaccine consists of synthesized sections of the CS protein. The difference is that the portions of the protein are fused with hepatitis B surface antigens and then mixed with adjuvants—chemicals known to boost the immune system. In 1997 Walter Reed reported that RTS,S had protected an unprecedented six out of seven volunteers in tests in Washington. Two years later SmithKline and Britain's Medical Research Council held full-scale field trials in The Gambia. Recruiting three hundred and six men from six rural villages, the MRC administered the vaccine in three separate doses. After two months, almost two thirds of the men were protected. Unfortunately, that rate fell to just 16 percent after fifteen weeks. It was too short a period to be of practical use, but for the first time, a vaccine had provided some protection against malaria in the field. Not only that, but the vaccine had worked in The Gambia—the place where many previous vaccines, including Patarroyo's, had met their Waterloo.

"The Gambia trial confirmed good and bad things about the vaccine," says Ballou cautiously. "Yes, it confers immunity on about half the people it is given to, but it lasts only a couple of months. Nobody believes that we have a vaccine that is ready for licensing or that will meet sub-Saharan Africa's needs yet."

Twenty or even ten years ago, such humility in malaria vaccine research was unheard of. But as immunologists have begun to realize the scale of the problem facing them, it is increasingly the norm. This is only partly due to the immunological challenges. In the laboratory, the parasite has been equally reluctant to give up its secrets. At the Sanger Institute in Cambridgeshire, where scientists have succeeded in mapping the human genome, to date only two of the parasite's fourteen chromosomes have been sequenced. The problem is that its genome is chock-full of adenines and thymines, two of the four nucleotide bases that are the building blocks of DNA. In some places these A's and T's average 80 percent, and on chromosomes 2 and 3 nearly 100 percent. "Falciparum has been far more difficult than any other organism we've tackled," says Sharon Bowman, the molecular biologist in charge of Sanger's falciparum-sequencing effort. "Because of the concentration of A's and T's, it's very difficult for us to clone out the bits of the parasite we need and then fit the pieces back together again." The real sticking point for the sequencers, however, has been a section known as the blob—three chromosomes that are so close in size they have to be treated as one.

In theory, as more of the pieces of the parasite become available, it should become easier to identify proteins for drug development and for splicing into a vaccine. However, many scientists are increasingly coming around to the view that DNA vaccines alone are not the answer. Instead, they are experimenting with what are known as prime-boost techniques—priming with a DNA vaccine and boosting the immune system with a second vaccine, either a recombinant protein or a recombinant virus. Hoffman, for instance, is now looking at boosting a DNA vaccine with RTS,S. Meanwhile, Adrian Hill, professor of human genetics at the Nuffield Department of Clinical Medicine in Oxford, has begun field trials in The Gambia of a six-part DNA vaccine inserted into an attenuated version of the vaccinia virus—the virus that eliminated smallpox. One of several prime-boost strategies being studied by Hill, the vaccine performed well in trials involving forty volunteers conducted in Oxford in 1999. Unlike Hoffman, Hill is not interested in the antibody response but is focusing exclusively on boosting T cells in the liver stage of the infection. But he does not rule out combining his techniques in the future with those of other researchers. "At the moment," he says, "the best vaccine is RTS,S. But at the end of the day we may find that we need to use several different approaches. My view is that malaria is so complex that for the moment, everyone has to concentrate on what they are best at. We should not expect any one person to come up with the answer. Those days of vaccine discovery are over."

One thing that would help is more money. In January 2001, the dangers of being a lone operator were dramatically brought home to Manuel Patarroyo when the San Juan de Dios Hospital Foundation—the parent organization of his Bogotá institute—collapsed, with debts of $150 million. Despite howls of protest from students, academics, and Nobel Prize–winning scientists and writers, the foundation's Basque bank seized Patarroyo's laboratory equipment as collateral and ejected his research team from their premises in the hospital's grounds. In May, following a series of court hearings—and the circulation of a ten-thousand-name petition—the bank agreed to return the equipment to Patarroyo, who in the meantime had set up a new organization, the Colombian Institute of Immunology Foundation, housed in a government building in another part of the city. According to Patarroyo, the disruption set back work on the new version of SPf66 by several months. "We expected to have the new vaccine ready by May 2002 at the latest. Now, it's difficult to say when we will be able to release it," he said.

Fortunately, scientists at universities and research institutions in North America and Europe enjoy greater security. But that does not mean that funding for malaria research is easy to come by. The United States currently spends $300 million a year on HIV and AIDS research. In contrast, it spends just $30 million on malaria—of which only about $5 million funds the military's work on the vaccine. In 1999, Bill Gates, the Microsoft computer billionaire, donated $50 million for vaccine development via a nonprofit organization established with his wife, Melinda, and his father, William. But while the money will accelerate research on vaccines already under way, it is a mere drop in the bucket set against the hundreds of millions of dollars poured by pharmaceutical companies each year into research on cures for more "profitable" diseases.

However, money is not the only problem. In the 1970s, the WHO succeeded in eradicating smallpox and came close to wiping out a host of other childhood diseases, such as polio and measles, by mobilizing hundreds of health workers across the globe. Today, the level of organization that would be required for a global malaria vaccine campaign simply does not exist. "We have to work in recognition that even if we had the best vaccine cheaply available, manufactured in large quantities, in our hand right now, we could not deliver it effectively to the children who are dying," admits Regina Rabinovich, the director of Gates's Malaria Vaccine Initiative.

In reality, we are nowhere near that point. Unless, of course, Patarroyo is right and his critics are wrong.

"I like to think of vaccines in two stages," says Hoffman. "First, we need a vaccine that provides temporary protection for nonimmune travelers and military personnel entering malarial areas. Second, we need a vaccine that protects children, growing up in areas where they are continually exposed to malaria, from dying. It could be that we come up with a vaccine that does both. The point is, it is a very complex puzzle. Whether we solve it this year, next year, or in ten to fifteen years' time, is what remains to be seen."

Epilogue

Warning—mozzies next 5 km.

IN THE TWO YEARS it took to research and write this book, mosquitoes and malaria have rarely been out of the news. No sooner did I begin looking into the legend of the Condesa de Chinchón, for instance, than it was reported that two Boy Scouts in Long Island, New York, had come down with the disease. According to Suffolk County health officials, the scouts had almost certainly been inoculated by mosquitoes breeding near a marsh on Long Island not far from where their camp was located. If so, it was the first case of locally acquired, as opposed to imported, malaria Suffolk County had seen in ten years.

A few weeks later, in September 1999, New York's mayor, Rudolph Giuliani, ordered the entire city to be doused with insecticide following the deaths of three adults from another mosquito-borne disease. This time the prime suspect was not the *Anopheles* but the *Culex* mosquito. At first scientists believed they were looking at an outbreak of St. Louis encephalitis, a rare avian virus normally confined to the southern United States. But within days investigators had established that the *Culex* were actually transmitting West Nile virus, an even rarer bird-to-human disease never before seen in North America.

The theme that linked these stories was global warming. Until the end of the nineteenth century, malaria was commonplace in the continental United States. From the Carolina swamplands all along the Mississippi to Illinois, and from Washington, D.C., as far west as Iowa, newspaper caricaturists poked fun at the "gaunt, lantern-jawed" inhabitants with their "careworn expression and languid mien." With the advent of federal dam and river projects, land clearance, and the popularization of quinine by medical purveyors like John Sappington, the disease—and the caricatures—gradually receded. But malaria has never been completely eliminated, and as the recent New York outbreaks show, it—or other exotic tropical diseases—could rebound at any time.

In Britain, because of its more temperate climate, malaria was always a more localized and seasonal occurrence. For the most part the disease was confined to the marshes and coasts of Kent, Essex, Norfolk, and Somerset, where it would flare up in summer and around harvest time before subsiding in winter. Nevertheless, by the early eighteenth century, field hands migrating north with the harvests had introduced malaria to the Scottish borders. There were even reports of "ague" in Aberdeen and Inverness.

As in North America, marsh reclamation reduced the anophelines' breeding areas, and the introduction of cinchona bark, and later quinine, reduced the pool of infectious parasites. But the key factor in Britain in reducing the incidence of malaria was probably the introduction of new root crops, like the turnip, that increased both the size and the health of animal herds and encouraged farmers to winter cattle in stables away from human habitations. *Anopheles atroparvus*, the main vector of malaria in Britain, began feeding on cows and horses instead of people, dramatically reducing the incidence of infections. If the British climate were to warm by just a few degrees, however, those same anophelines might begin transmitting vivax to people again—and possibly falciparum, too. All it would take is for one or two feverish backpackers returning from Asia to be bitten by a mosquito, and malaria could begin to take hold again in southern Britain—at least, that was the theory advanced by Britain's chief medical officer, Liam Donaldson, at a press conference in February 2001.

But global warming was not the only specter that stalked this book. While researching the history of the British cinchona missions, I could not help but notice that drug resistance was spreading. From Venezuela to Colombia, Brazil, Thailand, Kenya, and Tanzania, there were now strains of falciparum malaria that responded neither to chloroquine nor to quinine. Moreover, the tensions between profits and philanthropy first highlighted by Markham's bureaucratic battles with the India Office a hundred and fifty years ago had not gone away. If anything, they were even more pressing.

Until the end of the nineteenth century there was just one cure for malaria: quinine. The only question was how to bring it within reach of the masses. Today, no single drug is 100 percent effective against all strains of the disease, and developing a promising new antimalarial is an expensive proposition—at least $150 million spread over three years. Because of the rate at which resistance is spreading, some health economists argue that the West should be investing at least $500 million a year on new malaria drugs. Instead, we are spending only about $84 million. That translates into $42 per fatal case, significantly less than we spend on cancer or AIDS. Despite the fact that the G8 countries are now devoting substantial resources to malaria control—$750 million over the next ten years, under the WHO's Roll Back Malaria program—it is still not enough. Sooner or later, argue leading malaria experts, we will face a major health calamity.

When it comes, it is in Africa that the impact will be most acute. The World Health Organization estimates that at present some three thousand people die of malaria in sub-Saharan Africa a day for want of rapid and effective treatment. That is the equivalent of a small stadium full of people every week, or about 1.1 million a year—the majority of them children under the age of five. Because of global warming and the collapse of public health systems, however, those numbers are about to get a lot worse. In the mountains of western Kenya, for instance, malaria-carrying mosquitoes have been found breeding at six thousand feet for the first time in living memory, leading to an explosion of untreated cases. Doctors in Kisii, one of the worst-affected areas, blame abnormal rainfall patterns and the high cost of new antimalaria drugs.

Then there are the more frequent natural disasters, like the floods in Mozambique in March 2000, in which whole families were left stranded in trees, their bodies swarming with mosquitoes and other disease-carrying insects as they waited for the arrival of helicopters that could airlift them to safety. Those floods resulted in a fourfold increase in malaria in Mozambique. Similarly high rainfall in the northern highlands of Burundi in November 2000 resulted in a quarter of a million new cases and threatened even greater levels of infection in 2001. In the worst-case scenario, we are talking about five hundred million infections worldwide. That's half a billion new fever cases every year.

But morbidity is not the only measure of malaria's impact; it is also a huge drag on economic productivity. According to a recent study by Harvard University health economist Jeffrey Sachs, if the WHO had succeeded in its aim of eradicating the disease thirty-five years ago, Africa's gross domestic product would be about $100 billion greater than it is now. In the words of the WHO's director, Gro Harlem Brundtland, malaria is an "economic handicap" that not only hurts the living standards of Africans today but is "preventing the improvement of living standards for future generations."

One of the places where the issues of drug resistance, global warming, and the role of philanthropy in twenty-first-century malaria control increasingly come together is at the London School of Hygiene and Tropical Medicine (LSHTM). Founded in 1899, the school has long been at the forefront of research into malaria and other tropical diseases. It was here that in 1900, Sir Patrick Manson arranged for his son, Patrick Thorburn Manson, and a laboratory assistant to be bitten by a malaria-infected *Anopheles* forwarded from the Santo Spirito Hospital in Rome, thus proving that the disease was transmitted from mosquito to man. And it was here that researchers first began warning in the early 1980s of the risk of a resurgence in tropical diseases because of global warming and increasing drug resistance.

In recognition of the school's importance as a world center for malaria research and training, Bill and Melinda Gates recently donated $40 million to fund its work on new chemotherapy treatments, insecticides, and vaccines and to strengthen its existing links with scientists and health ministries in malaria-endemic regions. One possibility for future research is to use the latest satellite mapping technology to pick out areas of unusually high rainfall in Africa as a way to predict sudden outbreaks of the disease. Another is a project with Imperial College to produce the world's first genetically modified mosquito (the idea is to engineer a mosquito resistant to malaria that could invade wild mosquito populations). But the solution that offers one of the best chances of reducing morbidity and mortality in Africa is also one of the least complex: the mosquito net.

Long before the discovery that mosquitoes were the vectors of malaria, the mosquito net was an essential part of the explorer's kit. David Livingstone, for instance, slept under a "mosquito curtain" throughout his travels in Africa and was so thankful for the rest it brought him that he later remarked that its inventor "deserve[d] a statue in Westminster Abbey." In the summer of 1900 two researchers from LSHTM proved that a mosquito screen provided perfect protection from malaria when they spent four months shut up in a wooden hut on the Roman Campagna.[1] Today the WHO is just as wedded to the invention as Livingstone, seeing the provision of bed nets as a key component of Roll Back Malaria, which aims to cut mortality from malaria in Africa in half by 2010.

One of the campaign's leading supporters is Chris Curtis, LSHTM's professor of entomology. A short, dark-haired man with a permanently bemused expression, Curtis can usually be found showing graduate students around the school's vaults under Gower Street. There Curtis breeds some of the deadliest mosquitoes in the world: *Anopheles gambiae* from Africa, *A. stephensi* from Asia, and *A. albimanus*—the species that decimated French canal workers in Panama. The only clue that you are about to enter an insectary is a yellow triangle on the door that reads "Warning—mozzies next 5 km." That and the humid 80°F atmosphere.

According to Curtis, the office workers tramping to work along Gower Street have absolutely nothing to fear from the killers lurking beneath their feet. "None of the mosquitoes carry malaria," he says. Instead, Curtis, who is a leading authority on mosquito resistance to insecticides, is breeding the anophelines to find more efficient ways to kill them. One of the best methods, he believes, is bed nets dipped in pyrethroids—naturally occurring insecticides derived from the pyrethrum plant, which, like cinchona, is found at higher altitudes in Ecuador, Kenya, and Tanzania.

Studies show that bed nets treated with pyrethroids can reduce infant mortality from malaria by as much as half in malaria-endemic areas. If every child in Africa was provided with his or her own bed net, Curtis calculates, half a million lives would be saved each year.

There is just one problem. A bed net treated with a synthetic pyrethroid like permethrin costs $4. Even if you halve that through subsidies, it is still more than most Africans living in poor, rural areas of Tanzania, Kenya, Nigeria, and Congo can afford. Then there is the question of how to provide the chemicals to retreat the bed nets once the permethrin wears off.

The answer, according to Curtis, is to follow the example of governments such as that of Vietnam, which provide both the nets and the retreatments free of charge. But although studies show that the more people who have nets, the greater the protection from malaria for the community as a whole, few African nations can afford such largesse. The result is a compromise. Under the Roll Back Malaria program, governments are being urged to create more private-public partnerships to reduce the cost of nets, while increasing subsidies to the poor. The WHO's aim is that every African child have his or her own bed net within ten years. But Curtis worries that because bed nets are seen as a form of "personal protection," African governments will try to shuffle off responsibility to the very people who are most in need of help—subsistence farmers. "It's a vicious circle," says Curtis. "I hope the WHO's program succeeds, but my suspicion is that it will take a lot more donor money."

THERE IS no starker illustration of the imbalance between rich and poor than the wealth of antimosquito merchandise on display at camping stores in Europe and North America. At Black's in London, for instance, the shelves groan with mosquito nets, sprays, coils, and ever more sophisticated electronic devices for diverting anophelines from the "unsalted" blood of affluent Western tourists. That such protections are available for less than the cost of the air tax from Heathrow to Nairobi only makes it the more baffling that every year some two thousand people return to Britain infected with the parasites that spread malaria.

The point is that with the exception of parts of Thailand, Cambodia, Brazil, and other regions where drug resistance is growing, malaria is by and large an avoidable disease. All it takes is a few simple precautions. Before setting off for Venezuela to retrace Spruce's route along the Orinoco and Casiquiare in January 2000, for instance, I asked my GP for a course of mefloquine—a prophylactic that is still highly effective against most South American strains of the disease. I also amassed an impressive collection of sprays and creams and invested in a jungle-green permethrin-soaked net. The net, unfortunately, turned out to be quite useless—a Venezuelan entomologist laughed when she saw it, saying it was the sort of thing she used to trap mosquitoes. But the mefloquine protected me until I found a replacement, and steeped in Deet and my shop-bought repellents, I found the mosquitoes were more of a nuisance than anything else.

If I wanted proof of the consequences of ignoring such simple precautions, however, I did not have to look very far. At Tamatama, a Protestant mission post near Esmeralda—the village where Spruce was driven to distraction by "plagues" of mosquitoes in 1853—I discovered that scores of people had been hospitalized with falciparum malaria in the first week of January alone. As the missionaries admitted, they had no one to blame but themselves. Although malaria is widespread in the region, particularly among the Yanomami and other Indian communities living on the river, there had never been an outbreak at Tamatama before. In consequence, no one bothered with prophylactics. "Our sleeping quarters are screened in and the grounds are sprayed regularly, so we assumed it wasn't necessary," explained the mission nurse, Brenda Barkman.

However, just behind the compound lies a lagoon—the ideal breeding ground for anophelines. In the second week of December, when missionaries began arriving at Tamatama from all over the Orinoco to celebrate the last Christmas of the twentieth century, it was by the lagoon that many of their Indian charges chose to camp. Although no one can be sure, it was probably here that the malaria became established and began to spread to the rest of the post.

The decisive moment came on Christmas Eve, when the Indians, the missionaries, and their children—about a hundred people in all—gathered for dinner in the refectory. Unlike the barracks, there were no screens on the building, and as it was a special occasion, the festivities lasted well into the night.

The first case of falciparum malaria occurred five days later. By the end of the first week of the New Year, some seventy people had the disease. Barkman prescribed chloroquine and primaquine, then mefloquine. But nothing worked. It was only when she administered repeated courses of quinine that she at last saw the parasite loads begin to fall. It took five weeks for all her patients to recover. Afterward, Barkman ordered the refectory to be screened in and banned nighttime volleyball—until then a fixture of the missionaries' California-meets-jungle lifestyle. As far as I know, the curfew is still in force.

The missionaries were not the only casualties I encountered during the journey upriver. Two days after our stop at Tamatama, my Venezuelan guide, Hugo, also began to complain of headaches and fever. By the time we reached San Carlos on the Río Negro, he was sweating profusely. This time the diagnosis was vivax malaria. Although Hugo had slept under a mosquito net, like the missionaries he had failed to take a prophylactic.

Hugo's malaria eventually responded to chloroquine and proguanil. But it is only a matter of time before strains that are resistant to chloroquine, mefloquine, and quinine invade the Orinoco, too. What drugs will we look to then and who will provide them? One answer is WRAIR in Maryland, another is the WHO's Medicines for Malaria Venture. But as laudable as these organizations and initiatives are, ultimately they will count for little unless they win the backing of the pharmaceutical companies. Unfortunately, while the prophylactic malaria drug market is worth an estimated $335 million a year, Western pharmaceutical firms have little incentive to invest in research and development into new curative drugs. This is because in order to be affordable to poor people in developing countries, malaria drugs cannot be priced at more than $1 a dose, and should ideally cost about 2 cents a dose. In other words, to ensure that there is a demand for new treatments, pharmaceutical companies have to offer large subsidies, or simply be prepared to give the drugs away—as occurred with GlaxoSmithKline's Malarone donation program in Kenya in 1999.

It is not all doom and gloom, however. The free market has been harnessed in the cause of philanthropy before, and for concrete proof you need wander only a few blocks north of the LSHTM to the junction of Gower Street and the Euston Road. There, overlooking the traffic-clogged artery opposite Euston Station, are the offices of the Wellcome Trust—the largest medical research charity in the world. Each year the trust donates nearly $784 million to medical research, of which about 4 percent is spent on malaria. That the source of this money is the Wellcome Foundation—a business organization built on the profits of the pharmaceutical company Wellcome PLC (now known as GlaxoSmithKline)—is a tribute to the acumen and vision of one man.

HENRY SOLOMON WELLCOME WAS BORN in Garden City, Minnesota, in 1853. The son of a Methodist preacher, he is best remembered today for coining the term *tabloid* to describe the marketing, with his partner Silas Burroughs, of the world's first compressed medicines in 1880. The pills—which included such essentials as "tabloid" quinine sulphate—were an instant hit, and it was not long before everyone from kings to presidents to explorers was extolling their virtues. Tabloid quinine found its way into King Edward VII's automobile, the saddlebags of Theodore Roosevelt, and the supply pack of the African explorer Henry Morton Stanley, with the result that the London-based Burroughs-Wellcome became one of the most profitable pharmaceutical concerns of all time.

But Wellcome was not merely a brilliant businessman and publicist, he was also—like Sir Clements Markham—a humanitarian with a strong philanthropic streak. During a trip down the Nile in 1901 he had seen at first hand how malaria, smallpox, and famine had blighted the lives of the Sudanese following Kitchener's brutal suppression of the 1898 uprisings. As Wellcome later told a congressional committee, that experience convinced him that "in times of distress, the whole world is akin." Two years later he founded the Wellcome Tropical

Research Laboratories in Khartoum and instructed Andrew Balfour to supervise the draining and clearance of the *Anopheles* breeding grounds. Within a few years, mortality from malaria in and around Khartoum had been cut by 90 percent.

But to discover the roots of Wellcome's fascination with malaria, we have to go back much further, to his experiences as a traveling salesman in South America. It was this that gave birth to his lifelong empathy for the sufferings of indigenous peoples, and the catalyst, as for so many travelers before and since, was the legend of the Condesa de Chinchón.

After graduating from the Philadelphia College of Pharmacy in 1874, Wellcome went to work for a small company in New York. Then, in 1876, he was offered a trial with McKesson & Robbins, at the time one of America's largest pharmaceutical firms. His job was to promote the company's new gelatin-coated pills to doctors and druggists. The pay was just $16 a week, but in addition to taking him all over North America, the job offered him the opportunity to travel to South America and investigate local Indian remedies.

Concerned about the growing shortage of cinchona bark, in 1877 Wellcome's superiors decided he should visit Ecuador and report on the condition of the bark forests. In June, Wellcome and a companion landed at Guayaquil and caught a steam launch up a branch of the Guayaquil River to Pueblo Nueva. Wellcome had long been fascinated with the story of quinine's origins and was eager to question the Lojans about the legend of the condesa's cure. But first he had to climb through the cloud forests and cross over the freezing sierra.

In effect, Wellcome retraced Spruce's route in reverse, trekking around the summit of Chimborazo from west to east. It was there, just below the narrow-gauge railway that cuts across the sierra in the shadow of the volcano, that he saw his first cinchona tree. As he subsequently recalled in an article on his travels, he was picking his way down a slippery mud slope when "suddenly our guide shouted 'cascarilla' and we were gladdened by the sight of several fair-sized trees of *cinchona succirubra*."

Like Humboldt and Spruce, Wellcome was mesmerized by the bright green-and-red foliage and the way the leaves seemed to "glisten" in the harsh Andean light. Unfortunately, when he reached Loja he found that the hillsides had been stripped of mature trees and that the "outlook for bark supplies was exceedingly discouraging." Nevertheless, he used his time profitably, questioning the local inhabitants about the origins of the cure.

Wellcome was suspicious about the absence in the Spanish chronicles of any reference to the Indians' medicinal use of cinchona bark and pointed out that "it was the policy of the conquistadors to appropriate to themselves all creditable things." Speculating that the use of cinchona was probably confined to local tribes living near the bark forests, he concluded, "It is a general belief among the natives that cinchona bark was well known and highly regarded as a remedy by their ancestors long before the Spaniards, under the daring Pizarro, invaded their coast."

Wellcome's most lasting memory of Ecuador, however, was not of the cinchona tree but of the fevers and sufferings of the Indians. "The malaria," he wrote, ". . . is simply fearful, and owing to the great exposure and want of nutritious food the Indians yield very quickly to its influence." As if that were not bad enough, the gorges were strewn with the skeletons of Indians who, overloaded with cinchona bark, had lost their footing struggling to carry the valuable febrifuge out of the forests. "It is only by extreme poverty," he noted, "or obligation as peons, that they are induced to enter the bark forests to encounter the dangers for the meager pittance of ten to twenty-five cents per day."

Later on, in his travels in Africa, Wellcome experienced the scourge of other insect-transmitted diseases—notably leishmaniasis, sleeping sickness, and yellow fever. But malaria always remained a central concern. He twice succumbed to the disease, and, following a visit to Panama in 1910 on behalf of the U.S. War Department, he became a passionate supporter of Gorgas's sanitation campaign against mosquitoes.

It is little wonder then that on his death in 1936, Wellcome stipulated that his shares in Burroughs-Wellcome be used to fund medical research aimed at "the discovery, invention and improvement of medicinal agents . . . and the control or extermination of insects and other pests which afflict human beings and animal and plant life in tropical and other regions."

Today that bequest is worth approximately $21 billion.

For all his interest in medical research and medical history, however, Wellcome died convinced that Sebastiano Bado's story of the condesa's cure was true. It was only after his death that, ironically, a Wellcome medical museum researcher unearthed the conde's diary and other documents in Seville, which showed that the Condesa de Chinchón was almost certainly never treated for malaria in South America.

Fortunately, however, her legacy lives on—not just in the quinine still used to treat severe cases of falciparum malaria, but in medical research into new antimalarials funded, in part, by Henry Wellcome's slick marketing of her cure.

* * *

IN 1866, SIR CLEMENTS MARKHAM also decided to investigate Bado's story. Boarding an omnibus in Madrid, he traveled to Chinchón and questioned local villagers about the legend's provenance. To his gratification, he found that the story of how the condesa had distributed the bark on her husband's estates had passed into local folklore.

Markham later called for the Linnaean botanical nomenclature to be overturned, and for the correct spelling—"chinchona"—to take its place. But instead, delegates at the 1886 International Botanical Congress in London voted to retain Linnaeus's spelling on the grounds of "priority."

Peru never blamed Markham for the theft of the cinchona tree. Instead, in 1921 the Peruvian embassy in London presented the Royal Geographical Society with a bust of Markham in thanks for the support he had given Peru during its war with Chile. In his speech, Peru's chargé d'affaires praised Markham's role in the cinchona missions, saying they had rendered "a great service to humanity."

* * *

IN 1971 BOTANISTS from all over the world gathered at Coneysthorpe, in the shadow of Castle Howard, to unveil a plaque in honor of Spruce at the cottage where he spent the last seventeen years of his life. The inscription reads: "Richard Spruce of Ganthorpe, Welburn and Coneysthorpe: Distinguished botanist, fearless explorer, humble man."

* * *

IN 1986, ACF CHEMIEFARMA, a Dutch company that was then the leading manufacturer of quinine in the world, erected a new tombstone at Charles Ledger's grave in Rookwood Methodist Cemetery in Sydney. It reads: "Here rests Charles Ledger. He gave quinine to the world."

* * *

AS YET, there is no monument to Manuel Incra Mamani.

Afterword

It is not every day that the global war on malaria receives higher billing than the global war on terrorism but on October 2, 2002, a breakthrough in malaria research briefly knocked al-Qaeda and the ongoing crisis in the Middle East off the front pages. After years of analysis, scientists at three genome sequencing centers in the United States and Britain announced in *Nature* that they had finally cracked the genetic code of *Plasmodium falciparum*. Simultaneously, another group reported in *Science* that they had sequenced the genome of *Anopheles gambiae*, the deadly and efficient African vector of malaria. As the publications pointed out in separate editions devoted to the breakthroughs, this meant that for the first time in history biologists possessed the complete genetic makeup of the three organisms in malaria's life-cycle (the human genome having been sequenced two years earlier).

"This is an extraordinary moment in the history of science. At last, the enormous power of modern technology is penetrating the mysteries of an ancient disease," said Dr. Carlos Morel, director of the WHO's Tropical Diseases Research program.

"It will be a little while before the knowledge provided by the genome project is translated into practical tools," added Professor

Brian Greenwood of LSHTM more cautiously, "but this will happen and malaria will finally be brought under control."

Not everyone was prepared to be so categorical, however. *The New York Times'* headline read, "Genetic decoding *may* bring advances in worldwide fight against malaria" (my italics), while in a news feature headed with the question, "What difference does a genome make?" *Nature* stressed that because of regulatory bottlenecks, it might be as long as twenty years before the genomic information is converted into effective vaccines.

The undercurrent of skepticism running through these articles was understandable given malaria's history of false dawns. After all, hadn't we been here before in 1956 when the WHO (armed with chloroquine and DDT) pledged it would have malaria licked within a decade? It was only when the media's attention shifted back to that other war—normal service was soon resumed with the Bali bombing—and scientists were able to study the reports in more detail, that a more balanced assessment emerged.

The first point to make is that the codes *do* represent a giant leap forward. Previously, biologists have had to study the parasite gene by gene; now they can see all the strengths and weaknesses of the plasmodium at a glance—including every trick it uses to evade the human immune system. But deeper knowledge of the parasite's DNA has also brought renewed respect for the plasmodium's complexity. Though *falciparum* has far fewer genes than *Anopheles gambiae*—5,300 as opposed to 14,000—it took The Institute of Genomic Research (TIGR) in Rockville, Maryland, and their opposite numbers at the Wellcome Trust Sanger Institute in Hinxton, England, six years to sequence (*Anopheles*, by contrast, took just fifteen months). One of the first things biologists noticed was that the anchor proteins the merozoite uses to attach itself to the surface of red blood cells came in fifty different versions. This is not good news for vaccine designers: the anchor proteins or VAR genes encode the tiny protrusions on the surface of the remodeled red blood cells that are most closely associated with the human immune response. Take aim at one protrusion and the parasite can simply turn on another version of the VAR gene—incidentally, it is the protrusions that cause the blood cells to become sticky, clogging the capillaries to the brain (a pathology that in severe cases of *falciparum* can result in brain death).

As Malcolm Gardner, the scientist who lead the *falciparum* sequencing effort at TIGR, put it, "We already knew the parasite had several versions of the VAR gene but this is the first time we have been able to see its full repertoire. Knowing these variations means we are in a better position to target the enemy's defense system but this is by no means it for malaria."

The parasite's antigenic variation is not the only reason for caution. Although in July 2001 the G8 countries pledged an additional $1.8 billion to the Global Fund to Fight AIDS, TB, and Malaria, prevention and treatment programs remain woefully underfunded. Current worldwide donor spending, including the money pledged by the G8, is just $100 million to $200 million per year. By contrast, the needs of malaria research and development actually exceed $2 billion per year—probably more if donor agencies are to fund replacements for drugs like chloroquine and mefloquine (Lariam) to which resistance has developed.[1] The result is that just as the India Office had to wait fifteen years for its *Cinchona* plantations to start producing quinine, so the WHO may be waiting some time for the benefits of the new genomic technology to flow to the malaria frontline.

The irony is that this underfunding crisis coincides with a period of rapid technological advance. The problem facing malariologists today is that the choices are almost *too* great. Before the sequencing of *falciparum*, vaccine designers were already examining thirty promising protein candidates. Now, says Gardner, "they're drinking from a fire hose."

However, some immunologists argue the problem is not identifying new antigen targets but finding ways to make the immune system respond better to the targets we already know about. "If there were an extra $100 million to spend on malaria-vaccine research, I would allocate very little of it to exploring the parasite genome," argues Adrian Hill, a geneticist at the Nuffield Department of Medicine in Oxford. Instead, Hill would like to see scarce resources directed towards existing vaccines.

Similar cost-benefit arguments also apply to new drug therapies.

Biochemists had long suspected that *falciparum* used a particular enzyme to produce essential fatty acids. Analysis of the genome confirmed these enzymes were controlled by the apicoplast—a plant-like organelle that is a relic from the days when a primitive ancestor of the plasmodium fed on algae. Having identified the apicoplast's key role, researchers at the Justus-Liebig University in Glessen, Germany, looked for compounds that blocked its functions. Ironically, they discovered the most effective was a common herbicide and antibiotic, Fosmidomycin, already well known to pharmaceutical companies. Having tested well against malaria in the lab, Fosmidomycin is now entering phase two of clinical trials. But why go to all this technology and expense, argue advocates of traditional plant-based antimalarials, when there are plenty of herbal remedies out there that have still to be assessed? Take Changsan, a five-thousand-year-old Chinese remedy mentioned, like *Artemisia annua*, in ancient Chinese medical tests. Changsan's principle component is *Dichroa febrifuga*, a plant containing febrifugine. But though febrifugine, like Fosmidomycin, has been found to kill malaria parasites in the lab, Changsan has yet to be trialed in a clinical setting.

The list goes on. You can buy Neem, a traditional Ayurvedic antimalarial derived from the tree *Azadirachta indica*, on the Internet but not from your local physician, in spite of the fact that Mahatma Gandhi was an advocate. Ditto the traditional Malian remedy "Malarial," containing three herbs with proven antiparasitical activity.

To add to this embarrassment of riches, publication of *A. gambiae*'s genome has also brought closer the possibility of making it and other species of *Anopheles* malaria proof, or "refractory" to the disease. Researchers at Case Western Reserve University in Cleveland, Ohio, have shown that by inserting a synthetic gene into the related species, *A. stephensi*, they can almost completely disrupt the mosquito's ability to transmit *Plasmodium berghei*, the parasite species that causes malaria in mice. By tinkering with the genetic makeup of *A. gambiae* and inserting the resulting transgenic mosquitoes into wild populations, the hope is that they will similarly impede *A. gambiae*'s ability to transmit *falciparum* to humans. But such a strategy is fraught with ethical, legal, and public health issues. How can scientists ensure that the engineered mosquitoes will completely infiltrate wild populations? Would the genetic modification remain stable over time? And what if the malaria parasite becomes resistant to the transgenic mosquitoes just as it has to antimalarial drugs? A temporary halt in malaria transmission could result in people losing all natural immunity, rendering the disease even more devastating once the parasite returns. As Andrew Spielman, a medical entomologist at Harvard University, told *Nature*: "If malaria were eliminated from Africa for a short period of time, the consequences could be terrible."

But there is another consideration too. Even if the medical and ethical objections were overcome and the biotechnology was available and ready for use today, would African governments with health budgets of $5 per head per year be able to afford it? Probably not, says Chris Curtis, an entomologist at LHSTM who continues to argue the money would be better spent on low-tech interventions such as insecticide-treated bed nets. Unfortunately, this may not be a solution for much longer either. Pyrethroid resistance has already been reported in *A. gambiae* in West Africa and is now emerging in another malaria-carrying specie, *A. funestus*, in Mozambique. Moreover, despite the pledge to provide sixty percent of Africans at risk of malaria with a bed net by 2005, the WHO's Roll Back Malaria program is already falling well short of expectations. At the time of this writing, only seventeen of the forty-nine malaria-affected countries in Africa had lifted taxes on insecticide-impregnated nets—one of the key promises made at a malaria summit at Abuja, Nigeria, in 2001. The result is that in the majority of countries where people are most at risk, fewer than five percent of children have access to nets, while outdated drugs and insecticides continue to be the mainstay of frontline treatment.

The rich countries have not kept their end of the bargain either. In a hard-hitting attack on leaders in the developed world on the eve of a UN General Assembly session on malaria in November 2002, Jeffrey Sachs, a health economist and special adviser to the UN Secretary General, Kofi Annan, warned that the Global Fund to Fight AIDS, TB, and Malaria was on the verge of bankruptcy. The World Bank, which had promised $500 million to fight malaria, had "come nowhere near meeting this pledge" said Sachs, while among the donor states only Ireland had paid in full the money it had promised.[2]

It is not only the poor of Africa who are at risk. Weeks before the genome announcement, Americans awoke to the news that malaria mosquitoes had been breeding once again close to the nation's capital. The victims this time were a fifteen-year-old boy and a nineteen-year-old girl from Loudon County, Virginia. According to entomologists, the teenagers had probably contracted the malaria over the summer from mosquitoes in a park near the Potomac River, six miles from their homes. The significance was not lost on the genome scientists. In a symbolic nod to the threat of global warming, the strain of *falciparum* used in the sequencing effort had been cloned from a ten-year-old Dutch girl who had contracted malaria at Amsterdam's Schipol Airport in 1979.

The Virginia infections were not the only worrying news to emerge from the South that summer. In July three soldiers at an Army base at Fort Bragg, North Carolina, committed suicide after suffering dramatic mood swings following tours of duty in Afghanistan where they had been prescribed Lariam as a weekly prophylactic. Although U.S. Army epidemiologists and the wider scientific community downplayed the link, arguing that the soldiers' behavior could have had any number of causes, the incidents, widely reported in the media, highlighted concerns about past under-reporting of side effects associated with the drug.

It has always been known that mefloquine, like quinine, is a difficult drug to tolerate (as with quinine one of the adverse effects, for instance, is tinnitus). But anti-Lariam campaigners complain that in the past the drug's manufacturer, Hoffman-La Roche, downplayed both the incidence and persistence of more serious side effects such as anxiety and depression. Back in 1991, when Lariam was adopted as the standard prophylactic by the U.S. Army and the Peace Corps, scientists estimated the risk of "adverse reactions" as one in 10,000. And in 1995, Roche's production information warned only that side effects might last for "several weeks."

However, the latest study comparing Lariam with Malarone puts the risk as high as one in three.[3] Moreover, following a series of legal claims by patients, in August 2002 Roche adjusted its data sheets to acknowledge that psychiatric symptoms may persist "long after" Lariam has been stopped and that "rare cases of suicide ideation and suicide have been reported."

Given these reservations and the spread of drug-resistant strains of the parasite, it is little wonder that epidemiologists are once again looking to the Holy Grail of malaria research to eradicate the disease for good. GlaxoSmithKline's RTS,S vaccine is still considered the most promising candidate. Although the first field trial in The Gambia in 1999 provided protection for only two months, new versions are now being trialed in various settings. The largest is a double-blind study funded by the Malaria Vaccine Initiative in The Gambia that combines RTS,S with another type of vaccine developed by Hill's Oxford Group. Hill's aim is to use fragments of the modified vaccinia virus (MVA) to boost the protection conferred by RTS,S to up to two years, thereby providing a crucial "window of protection" for infants at greatest risk of early morbidity. But Hill is not counting on only one version of the vaccine—at Oxford clinical trials are also progressing on some twenty versions, including a DNA vaccine designed to induce potent T-cell responses against the liver stage of the malaria infection.

Good news is needed soon. All over Africa the disease is resurgent. In South Africa alone there were 432 deaths in 2001 compared to just fourteen in 1992. Meanwhile, in Sierra Leone, where peace has returned after years of civil war, malaria continues to claim the lives of more children than the fighting ever did.

Only an historian with a very short memory would bet on these statistics improving any time soon. As British researchers discovered last year when they examined DNA taken from the bones of children buried at an archaeological site in Rome, malaria was also a major cause of child mortality in 450 A.D.

Chances are the disease will be with us for some time yet.

Note on Sources

THE PRIMARY SOURCE for Spruce's life was his two-volume *Notes of a Botanist on the Amazon and the Andes*, edited by Alfred Russel Wallace (London: Macmillan, 1908). However, in some places Wallace omits important passages or skirts key details of Spruce's travels, and to fill these in I also consulted Spruce's original journals and letters at Kew Gardens. Similarly, for Spruce's journey over Chimborazo, I made extensive use of his *Report on the Expedition to Procure Seeds and Plants of the Cinchona Succirubra, or Red Bark Tree* to the undersecretary of state for India, January 3, 1862. For an overview of Spruce's life and contribution to botany, I also referred to M.R.D. Seaward and S.M.D. Fitzgerald, eds., *Richard Spruce (1817–1893): Botanist and Explorer* (Kew: Royal Botanic Gardens, 1986).

FOR LEDGER'S LIFE AND TRAVELS, I referred to Gabriele Gramiccia's *Life of Charles Ledger, 1818–1905: Alpacas and Quinine* (Basingstoke: Macmillan, 1988) and Ledger's voluminous correspondence with John Eliot Howard at the Royal Pharmaceutical Society of Great Britain. I also found Norman Taylor's *Cinchona in Java: The Story of Quinine* (New York: Greenberg, 1945) useful for the subsequent history of his *Cinchona ledgeriana* seeds.

MARKHAM PUBLISHED TWO BOOKS on the cinchona missions, *Travels in Peru and India While Superintending the Collection of Chinchona Plants and Seeds in South America and Their Introduction into India* (London: John Murray, 1862) and *Peruvian Bark: A Popular Account of the Introduction of Chinchona Cultivation into British*

India 1860–1880 (London: John Murray, 1880). In addition, he arranged for all correspondence on the introduction of the cinchona plant into India to be entered in the parliamentary record. These records are contained in the Blue Books published by Her Majesty's Stationery Office and covering the period 1852 to 1870. As in all Markham's works on cinchona, the titles adopt his preferred "ch" spelling. For the narrative of his expedition to Tambopata in Peru and an insight into his motives and fears, I also consulted his *Private Journals* and *Chinchona Notebooks I and II* (1859–60) at the Royal Geographical Society in London. For an overview of Markham's other career, as a geographer and patron of polar exploration, I referred to Albert H. Markham's *Life of Sir Clements Markham* (London: John Murray, 1917).

Humboldt's description of the cinchona tree and his reflections on the origin of the cure are contained in his paper *An Account of the Cinchona Forests of South America Drawn up During Five Years Residence and Travels on the South American Continent*, translated by Aylmer Bourke Lambert (London: Longman, 1821).

La Condamine's description of the cinchona tree, *Sur l'arbre du quinquina*, was published in the *Mémoires de l'Académie Royale des Sciences de Paris*, 1738, pp. 226–43 (Amsterdam: P. de Coup, ca. 1745). For his subsequent voyage with the seeds to Brazil, see his paper *Relation abregée d'un voyage fait dans l'interieur de l'Amérique Méridionale . . . en descendant la Rivière des Amazones*, read before the academy on April 28, 1745 (Paris: Veuve Pissot, 1745).

I found de Jussieu's description of the cinchona tree in *Description de l'arbre à quinquina: mémoire inedit de Joseph de Jussieu* (Paris: Societé du Traitement des Quinquinas, 1936).

Many scholars have tried to unravel the provenance of the story of the condesa's cure and the etymology of the word *quinaquina*. These questions are unlikely ever to be finally resolved, but I found the best forensic analyses were A. W. Haggis's "Fundamental Errors in the Early History of Cinchona" (*Bulletin of the History of Medicine* 10, 1942, pp. 586–92) and Jaime Jaramillo-Arango's *A Critical Review of the Basic Facts in the History of Cinchona* (Academic Press for the Linnaean Society of London, 1949, pp. 272–311). For a more recent analysis, I also found useful Fernando Ortiz-Crespo's *Fragoso, Monardes and pre-Chinchonian Knowledge of Cinchona (Archives of Natural History* 22 [2], 1995, pp. 169–81). The information on the Kina Bureau and quinine pricing between the wars was taken from the United States Department of Commerce's Trade Information Bulletin, *Quinine Production and Marketing* (no. 273, October 1924).

Notes

1. THE FEVER

1. The consequences are also potentially far graver, as the sand flies transmit leishmaniasis, a parasitic skin disease.

2. Or as one British sailing directory put it as late as 1883, "poisonous exhalations from the land." The canard was only finally laid to rest in 1897, when the British army surgeon Sir Ronald Ross succeeded in showing that malaria was caused by a parasite harbored in the mosquito's salivary gland and injected into its human host each time the insect took a blood meal.

3. Neither Spruce nor Wallace record any further details of the encounter. In the edited account of his fever in his *Narrative of Travels on the Amazon and Río Negro* (London and New York: Ward Lock, 1890), Wallace does not say whether he had the quinine with him, writing simply: "The ague, however, now left me, and in another week, as I could walk with a stick down to the river-side, I went to São Gabriel, to see Mr Spruce, who had arrived there, *and kindly been to see me a short time before* [my italics]." However, the passage leaves open the possibility that Spruce was the provider.

4. The exact date of Wallace's revelation is not known. It is thought that the theory came to him in the Spice Islands in either February or March 1858, but the precise date is lost because Wallace's diary is vague (possibly because he was confused by fever). As soon as he had recovered, however, he sent a four-thousand-word synopsis to Darwin, who received the letter on June 18, 1858. Wallace's ideas were so similar to his own as yet unpublished *On the Origin of Species* that two weeks later Darwin hastily submitted their ideas in a joint paper to the Linnaean Society in London.

5. Humboldt was also the first to note the way in which the different species of biting flies on the Orinoco stagger their attacks. "What appeared to us very remarkable, and is a fact well known to the missionaries, is that at different hours of the day you are stung by different species. Every time the scene changes and . . . other insects 'mount guard,' you have a few minutes, often a quarter of an hour, of repose. The insects that disappear have not their places instantly supplied by their successors." (Humboldt and Bonpland, p. 276.)

6. It is a pity the mosquitoes undermined Spruce's determination to climb the Cerro Duida. When the mountain was finally explored by two Americans in 1928 it yielded no fewer than seven hundred different plants, of which over two hundred were new to science.

7. The road, built in 1924 by the Venezuelan dictator Juan Vicente Gómez, is itself a bleak reminder of the fevers that plague this region. Hundreds of workers died from malaria during its construction, and it is said that their souls still haunt the area.

8. To escape the mosquitoes and press his botanical specimens in peace, Humboldt's traveling companion, Aimé Bonpland, was forced to crawl into a *hornito*, a smoke-filled oven without doors or windows. Humboldt writes that "the absence of the mosquitoes is purchased dearly enough by the excessive heat of the stagnated air, and the smoke of the torch of copal, which lights the oven during your stay in." (Humboldt and Bonpland, p. 281.) In fact, Bonpland probably already had malaria. Humboldt writes that he had been feeling lethargic ever since coming down to Maipures from San Fernando and that by the time they reached Angostura (Ciudad Bolívar) a few weeks later, he was seriously ill. After taking quinine and angostura bitters, however, he recovered.

9. Before the identification of malaria with specific plasmodia, the disease was defined by the interval between attacks: vivax was known as benign tertian because of the relative mildness of the paroxysms and because the fever occurred every second day, with the remission coming on the third day; *malariae* was known as quartan because the attacks occurred every fourth day; and falciparum was known as malignant tertian or pernicious fever because of the severity of the attacks, lasting anywhere from twenty-four to thirty-six hours, with only about twelve hours of intermission.

10. The hammock can still be seen in a display case at Castle Howard.

2. THE CURE

1. Linnaeus named the genus *Cinchona* in his *Genera Plantarum* of 1742. One theory is that in so doing he relied on Sebastiano Bado, the Italian physician who first popularized the legend of the condesa's cure in his 1663 book and who used the spelling *cinchon*, the Spanish "chi" sound being "ci" in Italian. However, although "cinchona" is now the accepted botanical nomenclature, confusion still reigns over the correct pronunciation, with English-speaking botanists saying "sin-kona" or "sin-chona" and Spanish speakers saying "cheen-chona."

2. The best barks were reserved for the exclusive use of the royal pharmacy at Madrid, hence the appellation "Crown" barks, while other, less valuable,

barks were sold at auction for export overseas. Crown barks originally sold at Panama for five and six dollars, and at Seville for twelve dollars, a pound, but they were later adulterated with less valuable barks, and the price fell to one dollar a pound.

3. In another, even more fanciful version of the legend, which first achieved popularity in 1817 as a play entitled *Zuma ou la Découverte du Quinquina*, by Mme de Genlis, a governess in the Duke of Orleans's household, it is the condesa's beautiful Indian servant girl, Zuma, who imparts the secret to her mistress. The most salient feature of this version is Zuma's reluctance at first to share the bark's miraculous healing powers with the hated conquistadors. Indeed, she has been sworn not to, the Indians taking great pleasure in watching their Spanish oppressors die from fevers to which they alone have the cure. It is only when Zuma is consumed by the same fever and her husband smuggles some of the precious bark into the palace at Lima to treat her that she takes pity on the condesa and cures her too. Ironically, in Mme de Genlis's play the court physicians suspect Zuma of poisoning her mistress and condemn her to be burned at the stake. Luckily, the condesa recovers in the nick of time, pardons Zuma, and the caciques come forward to reveal the virtues of the bark to the Spaniards.

4. Writing in 1663, Gaspar Caldera de Heredia, a Portuguese trader based in Seville, claimed that cinchona was first used by Indian miners in Ecuador for the relief of cold. To reach the rock face, the miners had to wade through freezing water, and they alleviated their "*rigor et frigor*" by taking pulverized cinchona bark. Observing this, Heredia says, the Jesuits requested samples of the bark and, applying logic, decided to test whether it was also beneficial in the cold and shivering of intermittent fevers.

5. In yet another version of the legend it was not the corregidor of Loja but the viceroy's physician, Dr. Juan de Vega, who suggested the condesa take the cure and who, on her recovery and return to Spain in 1640, brought a quantity of the bark with him, selling it at Seville, a then very malarious area of Spain, for an English sovereign an ounce (about £75 an ounce in today's money). In fact, as scholars have since shown, de Vega never left Peru but continued to practice medicine in Lima until at least 1650.

6. The problem with this argument is that if the cure was known to the Inca, why don't we find similar reverence for cinchona as for the coca leaf? The earliest finds suggest that coca was used on the coast of Peru by 2000 B.C. Remains of coca leaves have been found at Indian sites dating back to A.D. 600, and lime pots and ceramic figurines depicting people chewing leaves have been found at virtually every major site from every era of pre-Columbian civilization. We also know that for the Incas the coca leaf was a living manifestation of the divine. Coca fueled the construction of the Incas' monumental stone cities, kept their armies on the march, and enabled the imperial runners, the *chasquis*, to relay messages over distances of four thousand miles in a week. Perhaps the apparent lack of reverence for the bitter-tasting cinchona bark simply reflects the fact that the Incas were a highland people, and that before the Spanish conquest malaria was not a problem.

7. See John Thurloe, *A Collection of State Papers of John Thurloe, Esq.*, ed. Th. Birch, 7 vols. (London: Woodward and Davis, 1742, 4:18).

8. For a detailed account of Cromwell's symptoms and an analysis of his postmortem, see L. J. Bruce-Chwatt, "Oliver Cromwell's Medical History," *Transactions and Studies of the College of Physicians of Philadelphia*, 5 ser., v. 4, 1982.

9. The key foreign ingredient, of course, was quinine bark, but what was the other? One intriguing possibility is that Talbor added the leaves of sweet wormwood, *Artemisia annua*, a powerful antimalarial recently "rediscovered" in China, where it is known as *quinghaosu*. For a fascinating discussion of Talbor's remedy for malaria, see Mary Dobson, "Bitter-sweet Solutions for Malaria: Exploring Natural Remedies from the Past," *Parassitologia* 40 (1998): 69–81. According to Dobson, Talbor may have used wormwood to offset the bitterness of the cinchona.

10. Unfortunately, Sydenham rather spoils the effect of this endorsement in the next sentence, when, in what is clearly a swipe at Talbor, he continues: "which the physicians of London not being pleased to take notice of in my book, or not believing me, have given an opportunity to a fellow that was but an apothecary's man, to go away with all the practice on Agues, by which he has gotten an estate in two months, and brought great reproach on the faculty."

11. It is also said that Torti was the first to refer to intermittent fevers as *mal'aria*, from the Italian for "bad air." But the term never appears in his writings and appears to have already been in use in Italy more than a century before his birth. See Saul Jarcho, *Quinine's Predecessors*.

3. THE CURSE

1. The nickname "Old Grog" came from Vernon's habit of wearing a grosgrain cape in foul weather. But it was because of his introduction of rum diluted with four parts water that the word *grog* passed into the English language as a synonym for rum. In an effort to curb drunkenness at sea, Vernon replaced the standard issue of neat rum with the diluted mixture and made the issue take place twice a day rather than once. He thus hoped to avoid the accidents that occurred when men drank half a pint of neat rum in one go.

2. To ensure that the men would not take advantage of the directive and get drunk, the Admiralty stipulated that the medicated chests be kept under lock and key below deck, as was the practice with other spirits, and brought up only when a ship entered malarious waters or sailors went ashore.

3. Indeed, it was the first time Linnaeus had seen the flowers of the cinchona tree, and he wrote immediately to Mutis to thank him. "I received in due time, eight days ago, your letter dated 24th day of September of 1764, and was greatly moved and overjoyed by it, as it contained a beautiful drawing of the quina bark, together with leaves and flowers, which flowers, never seen by me until now, really have given me an idea of a very rare genus, and very different from that which I acquired through the figure of Monsieur Condamine. I am very grateful for everything." (Jaramillo-Arango, *Basic Facts in the History of Cinchona*, p. 302.)

4. In fact, as subsequent analysis has confirmed, red bark contains four alkaloids in near equal amounts: quinine, quinidine, cinchonine, and cinchonidine, while yellow bark contains much quinine and a little cinchonine. Gray bark contains cinchonine and tannin.

5. These are the presence of curly hairs bordering the laciniae of the corolla, the characteristic upward dehiscence of the capsule, and the little pits at the axils of the veins on the undersides of the leaves.

4. THE VISION

1. See Weddell, *Histoire Naturelle des Quinquinas.*

2. In 1865 Ledger overheard Henriquez, the Indian who had assisted the Dutchman, boasting how he had mixed arsenic with the earth in the Wardian cases and in the pots that Hasskarl used for watering his cinchonas. "Lucky for Hasskarl," writes Ledger, "that they did not poison him, as they could have done."

5. THE PHILANTHROPIST

1. Ruiz and Pavon named it *Cinchona uritusinga* after the mountain near Loja on which it was most abundant. In fact, it was identical to the variety sketched by La Condamine in the forests of nearby Cajanuma and named *C. officinalis* by Linnaeus. Markham later proposed that *uritusinga* should be known as *condaminea* to avoid confusion. However, his suggestion was not adopted.

2. "It is a curious coincidence that at the very time when Dr Royle was writing this report I was actually exploring some of the chinchona forests of Peru," Markham later recalled in his popular account of the cinchona missions, published in 1880. "But the object of my travels was of an antiquarian and ethnological character, and I was in ignorance of the desire of the Indian government to procure supplies of those plants, which I then only admired for their beauty." (*Peruvian Bark*, p. 88.)

3. Livingstone's experiments with quinine resulted in his formulation of the famous Livingstone or Zambezi Rousers, consisting of eight grains (520 mg) each of resin of julap and calomel mixed with four grains each of quinine and rhubarb. The pill's name was inspired by his remark that "two ordinary sized pills would be a rousing dose for a woman."

4. Despite having been administered quinine, 354 soldiers were admitted to Varna general hospital suffering intermittent fevers between June and August 1854, and more than 100 died. But considering that as many as thirty thousand soldiers had been camped in the malarious Danube delta since January, the casualty figures were low, particularly when compared with the number of deaths from other diseases, such as cholera and typhus.

5. Before setting sail from England in 1849, Spruce had seen a huge quantity of bark from these very forests lying unsold in a warehouse in Liverpool. The bark had been exported to England by a young Peruvian businessman, Don Luis Lopez, in the belief that it was identical to the *cascarilla roja* of Huánuco and that he was about to make his fortune. But when the bark arrived at Liverpool after an epic journey down the Amazon and across the Atlantic, it turned out that it was a false quina, *Condaminea corymbosa*, with no febrifugal powers whatsoever. "There had not been wanting people on the spot who warned Don Luis of his mistake; but he was too opinionated to listen to them, and persevered to his disastrous overthrow," Spruce wrote to Bentham.

6. THE QUEST

1. See *Peruvian Bark*, pp. 266–67.
2. Every time you purchase goods in Peru you are honoring this legacy, for the national coat of arms, complete with cinchona tree, appears on the reverse side of the Peruvian *sol* (silver dollar).
3. Don B. Vicuna Mackenna in a review of *Cusco and Lima* published February 4, 1860, in the *Commercio de Lima.*
4. The doctors treated the "stricture" with enemas and opiates, but the years of medical neglect had taken their toll, and for the remainder of his life Spruce was unable to sit up at a microscope without the risk of internal bleeding and could only write by reclining in a chair with his papers spread on his lap.

7. SPRUCE'S RIDGE

1. The boy's fever subsequently fell and he was taken to Guanujo. Spruce never learned what became of him but writes that "marsh fevers are generally cured, as if by magic, by simply removing the patient to the cool rarefied atmosphere of the Andes." (*Report on the Expedition to Procure Seeds and Plants of the Cinchona Succirubra* . . . , p. 75.)
2. Markham later wrote: "If I had received any proposal from him I should, without hesitation, have taken upon myself the responsibility of engaging Mr Ledger's services, and he would then certainly have received some, though probably inadequate, remuneration." (*Peruvian Bark*, p. 214.)
3. Ledger later wrote that although he could ill afford to help Markham, he "could not resist the temptation of obtaining what was so much wanted."

8. THE SHADOW IN THE FOREST

1. The Royal Pharmaceutical Society notation identifies it as a picture of Santiago, but Gabrielle Gramiccia, who researched Ledger's life extensively, believed it to be a picture of Manuel. Either way, the picture is a graphic illustration of the confusion surrounding Mamani's life and identity. See Gramiccia.

9. A TALE OF TWO SEEDS

1. Nine years later, in 1873, Nightingale presented an almost identical paper, entitled "Life or Death in India," to the National Association for the Promotion of Social Science, in Norwich.
2. McIvor calculated that Money's pouch of *Cinchona ledgeriana* contained some twenty thousand seeds. If he had followed the same procedures as Van Gorkum, in theory he could have raised four million plants. In the end, however, only sixty thousand seedlings were planted, and only a small proportion of these were forwarded to Sikkim.
3. One hundred years later the Institute Pasteur would broadly vindicate the results. In in vitro studies in 1987 on plasmodium falciparum, using the most

up-to-date techniques, the institute found that, depending on the strain tested, quinidine and cinchonine were between five and ten times more active than quinine, and that a combination of all the alkaloids was between two to ten times more effective against strains that showed marked resistance to quinine.

4. A "bougie" is "a thin flexible surgical instrument made of waxed linen, india-rubber, metal, etc., for introduction into the passages of the body, for the purpose of exploration, dilatation, or medication." (Oxford English Dictionary.)

10. THE GREAT WHITE HOPE

1. Why Laveran observed exflagellation when everyone else had missed it is not clear. He was not the first to examine live blood from a malaria patient, but he may have been the first to wait fifteen minutes before doing so. This is just long enough to permit the male gametocytes to transform themselves into gametes, a process that usually takes place inside the stomach wall of the mosquito.

2. The reason is that the mature gametocytes do not ingest hemoglobin, so there is no toxic waste process for quinine to interfere with.

3. For these reasons quinine is considered a suppressive, as opposed to a "causal," prophylactic.

4. It was not a new idea. From the earliest days of the British raj, British officers stationed in India had dissolved quinine in carbonated water. Inspired by their example, Schweppes began manufacturing Indian tonic water in 1870. Other quinine-flavored drinks soon followed, including bitter lemon, and in France, Dubonnet. The practice eventually led to the invention of gin and tonic—one of the most popular drinks of all time, although of course the amount of quinine in the tonic water today is minimal.

5. There were ten European manufacturers of quinine: Vereinigte Chininfabrieken, Zimmer & Co., Frankfurt; C. F. Boehrigner und Sohne, Mannheim; Chininfabrik Braunschweig, Büchler & Co., Brunswick; Amsterdamsche Chininefabriek, Amsterdam; N. V. Nederlandsche Kininefabriek, Maarsen; Howard & Sons Ltd., London; Charles Buchet et Cie, Pharmacie Centrale de France, Paris; Pointet et Girard, Paris; Société du Traitement des Quinquinas, Paris; A. Taillandier Argenteuil (Seine-et-Oise), France. After 1892, however, they also faced competition from the Bandoengsche Kininefabriek plant in Java.

6. A unit, equivalent to five grams of quinine (1 percent of quinine per one-half kilo of bark), was the standard measure for pricing bark. In 1920s prices, one Dutch cent was equivalent to 0.46 U.S. cents.

11. THE MALARIA FIGHTERS

1. The problem was that far too much bark was still being produced. To reduce the supply and help maintain an economically viable price, the Association of Bark Producers in Java began a campaign in 1929 to persuade planters to retire from the business, with a view to then uprooting their plantations.

2. See Hehir, p. 291: "It is clear that the world could not at present meet this, the Indian demand, and also that the demand will never be realised while present world prices continue."

3. Notwithstanding this agreement, the Netherlands Indies government issued a series of ordinances six years later that further restricted bark exports.

4. It is not surprising that Perkin failed. When he began his work on quinine synthesis, its formula—$C_{20}H_{24}N_2O_2$—had only recently been settled, and the structure of the quinine molecule was unknown. The first person to unravel the correct connectivity between the molecule's atoms was the German chemist Paul Rabe, in 1918. But Rabe only carried out a partial synthesis, and it was not until 1944 that the Harvard chemist Robert Woodward picked up where Rabe had left off. In May 1944, Woodward and his student colleague, William von Eggers Doering, announced that they had synthesized homomeroquinene, which in turn could be converted into quinotoxine—the molecule from which Rabe claimed he had succeeded in obtaining quinine. With the United States cut off from the Dutch East Indies and the world's major source of cinchona bark, the breakthrough was hailed by the *New York Times* as "one of the greatest scientific achievements in a century." However, it took them fifteen steps just to reach quinotoxine, and they never went on to full synthesis of quinine, relying on Rabe's prior work for their "formal" proof. Unfortunately, as Gilbert Stork, professor of chemistry at Columbia University, showed in December 2000, when he came to carry out the first full synthesis of quinine, Rabe's proof was flawed. Quinine could not be reached from quinotoxine but only from another molecule, deoxyquinine.

5. Unfortunately, while Atabrine was a success against falciparum malaria, it merely suppressed the vivax form of the disease. Hence, when troops were put on malaria-free islands to recover and Atabrine was stopped, those infected with vivax relapsed.

6. Recently, however, DDT has been making a comeback. In December 2000 the United Nations approved limited use of DDT to control mosquitoes despite calls for a worldwide ban on the pesticide by Greenpeace and the World Wildlife Fund. The UN's decision followed intense lobbying by South Africa and the Malaria Foundation International, which argue that spraying houses with DDT is still one of the most effective and affordable methods of malaria control.

7. From the military view, a key advantage of chloroquine, and later the "CP" tablet, was that it only had to be taken once a week, meaning that it could be administered as easily in combat as in garrison. One of the reasons for the "failure" of quinine and Atabrine was that the pills had to be taken daily—a discipline hard to enforce in the field.

8. Novartis is marketing yet another artemisinin derivative, artemether, in combination with lumefantrine, while the WHO and the U.K. Department of International Development, in partnership with GlaxoSmithKline, are conducting trials in Africa with a chlorproguanil-dapsone combination called Lapdap.

12. THE VACCINE HUNTERS

1. Biologists have found a similar correlation between malaria and the incidence among Polynesian islanders and people of southern Mediterranean descent of thalassemia, another inherited recessive trait that results in a failure to manufacture hemoglobin. However, while thalassemia provides some protection against malaria, biologists are divided over whether its prevalence is due to selection pressure exerted by malaria.

EPILOGUE

1. The physicians G. Carmichael Low and L. Sambon and the artist A. Terzi erected the hut at Castel Fusano, near Ostia, a notoriously malarial area. They lived there from July through October 1900, venturing outside during the day but making sure they were always back inside before sunset. Their only view of the night sky was through a small wire-screen gauze. They took no quinine, and at the end of four months, when they were found to be free of malaria, the experiment was declared a success.

AFTERWORD

1. Jeffrey D. Sachs, "A New Global Effort to Control Malaria," *Science*, October 2002.
2. *The Financial Times*, "Battle with Malaria Losing Ground as AIDS Drains Funding," November 5, 2002.
3. Overbosch et al. "Atovaquone-Proguanil Versus Mefloquine for Malaria Prophylaxis in Non-Immune Travelers: A Randomized, Double-Blind Study," *Clinical Infectious Diseases*, October 2001, 33: 1015–1021.

Bibliography

BOTANICAL EXPLORATION

Coats, Alice M. *The Quest for Plants: A History of Botanical Explorers*. London: Studio Vista, 1969.

Cutright, P. R. *The Great Naturalists Explore South America*. New York: Macmillan, 1940.

Davis, Wade. *One River: Explorations and Discoveries in the Amazon Rain Forest*. New York: Simon and Schuster, 1996.

Furneaux, Robin. *The Amazon: The Story of a Great River*. London: Hamish Hamilton, 1964.

Hagen, Victor Wolfgang von. *South America Called Them: Explorations of the Great Naturalists*. London: The Scientific Book Club, 1949.

Humboldt, Alexander von, and Aimé Bonpland. *Personal Narrative of Travels to the Equinoctial Regions of America, During the Years 1799–1804*. 3 vols. London: Henry G. Bohn, 1852–53.

Musgrave, T., C. Gardner, and W. Musgrave. *The Plant Hunters: Two Hundred Years of Adventure and Discovery Around the World*. London: Orion, 1998.

CINCHONA AND QUININE

Acosta Solis, Misael. *La Cinchona, planta nacional del Ecuador*. Quito: Impr. del Ministerio del Tesoro, 1950.

Andersson, Lennart. "A Revision of the Genus Cinchona (Rubiaceae-Cinchoneae)." *Memoirs of the New York Botanical Garden* 80, January 30, 1998.

Bado, Sebastiano. *Anastasis Corticis Peruviae Seu Chinae Chinae Defensio.* Genoa: P. J. Calenzani, 1663.

De Vrij, J. E. *On the Cultivation of Quinine in Java and British India.* Translated by Clements Markham. London: Eyre and Spottiswoode, 1865.

Duran-Reynals, Marie Louise de Ayala. *The Fever Bark Tree: The Pageant of Quinine.* New York: Doubleday, 1946.

Hobhouse, Henry. *Seeds of Change: Six Plants that Transformed Mankind.* London: Macmillan, 1999.

Hodge, W. H. "Wartime Cinchona Procurement in Latin America." *Economic Botany* 2, July–September 1948.

Howard, John Eliot. *Illustrations of the Nueva Quinologia of Pavon.* London: L. Reeve, 1862.

———. *The Quinology of the East Indian Plantations.* 2 vols. London: L. Reeve, 1869–76.

Jarcho, Saul. *Quinine's Predecessor: Francesco Torti and the Early History of Cinchona.* Baltimore: Johns Hopkins University Press, c. 1993.

Lambert, Aylmer Bourke. *An Illustration of the Genus Cinchona: Comprising Descriptions of All the Official Peruvian Barks, Including Several New Species.* London: J. Searle, 1821.

Markham, Clements R. *A Memoir of the Lady Ana de Osorio: Countess of Chinchón and Vice-Queen of Peru, 1629–39, with a Plea for the Correct Spelling of the Chinchona Genus.* London: Troubner, 1874.

Mutis, José Celestino. *Expedición Botánica.* Colombia: El Ancora Editores, 1983.

Rainey, Froelich. "Quinine Hunters in Ecuador." *National Geographic*, March 1946.

Ruiz, Hippolito. *Travels of Ruiz, Pavon and Dombey in Peru and Chile, 1777–1788.* Translated by Bruce Dahlgren. Chicago: Field Museum of Natural History, 1940.

Taylor, Norman. *Quinine: The Story of Cinchona.* New York: Cinchona Products Institute, 1943.

Van Gorkum, Karel Wessel. *A Handbook of Cinchona Culture.* Translated by B. D. Jackson. London: Trübner, 1883.

Various authors. *Three Centuries of Cinchona: Proceedings of the Celebration of the Three Hundredth Anniversary of the First Recognized Use of Cinchona Held at the Missouri Botanical Garden, St. Louis, October 31–November 1, 1930.* St. Louis: Missouri Botanical Garden, 1931.

Weddell, Hughes Algernon. *Histoire Naturelle des Quinquinas.* Paris: V. Masson, 1849.

———. *Voyage dans le nord de Bolivie et dans les voisins de Pérou.* Paris: P. Bertrand, 1853.

MALARIA: SCIENCE, HISTORY, AND THE MILITARY

Bartlett, Elisha. *The History, Diagnosis, and Treatment of the Fevers of the United States.* Philadelphia: Lea and Blanchard, 1847.

Bayne-Jones, Stanhope. *The Evolution of Preventive Medicine in the United States Army, 1607–1939.* Washington, D.C.: Office of the Surgeon General, Department of the Army, 1968.

Blanco, Richard L. *Wellington's Surgeon General: Sir James McGrigor.* Durham, N.C.: Duke University Press, 1974.

Boyce, Sir Rupert W. *Mosquito or Man? The Conquest of the Tropical World.* London: John Murray, 1909.

Bruce-Chwatt, Leonard Jan. *Essential Malariology.* 2d ed. London: William Heinemann Medical Books, 1985.

Bruce-Chwatt, Leonard Jan, and Julian de Zulueta. *The Rise and Fall of Malaria in Europe: A Historico-Epidemiological Study.* Oxford and New York: Oxford University Press, 1980.

Cantlie, Neil. *A History of the Army Medical Department.* 2 vols. Edinburgh: Churchill Livingstone, 1974.

Celli, Angelo. *Malaria According to the New Researches.* Translated by John J. Eyre. London: Longmans, Green and Co., 1900.

Cuthbert, Christy. *Mosquitos and Malaria: A Summary of Knowledge on the Subject Up to Date.* London: Sampson Low, Marston and Co., 1900.

Desowitz, Robert S. *The Malaria Capers: Tales of Parasites and People.* New York: W. W. Norton, 1991.

Dobson, Mary. " 'Marsh Fever'—The Geography of Malaria in England." *Journal of Historical Geography* 1980, pp. 357–89.

Drake, Daniel. *Malaria in the Interior Valley of North America.* Urbana, Ill.: University of Illinois Press, 1964.

Falls, Cyril, and A. F. Becke. *History of the Great War.* 2 vols. London: Her Majesty's Stationery Office, 1933–35.

Harrison, Gordon. *Mosquitoes, Malaria and Man: A History of the Hostilities Since 1880.* London: John Murray, 1978.

Harrison, Mark. "Medicine and the Culture of Command: The Case of Malaria Control in the British Army During the Two World Wars." *Medical History* 40, 1996, pp. 437–52.

Hehir, Sir Patrick. *Malaria in India.* London: Oxford University Press, 1927.

Jaramillo-Arango, Jaime. *The Conquest of Malaria.* London: Heinemann, 1950.

Joy, Robert J.T. "Malaria in American Troops in the South and Southwest Pacific in World War II." *Medical History* 43, 1999, pp. 192–207.

Lemaitre, Eduardo. *A Brief History of Cartagena.* Colombia: Editorial Colina, 1994.

Lloyd, C., and J.L.S. Coulter. *Medicine and the Navy*, vol. 3: *1745–1815.* London: Livingston, 1961.

MacPherson, W. G., and T. J. Mitchell. *History of the Great War, Medical Services General History*, vol. 4. London: Her Majesty's Stationery Office, 1924.

Nye, Edwin, and Mary Gibson. *Ronald Ross: Malariologist and Polymath*. London: Macmillan, 1997.

Russell, Paul F. *Man's Mastery of Malaria*. London and New York: Oxford University Press, 1955.

———. *Practical Malariology*. 2d ed. London: Oxford University Press, 1963.

Shepherd, John. *The Crimean Doctors: A History of the British Medical Services in the Crimean War*, vols. 1 and 2. Liverpool: Liverpool University Press, 1991.

Spurgeon, Neel. *Vietnam Studies: Medical Support of the U.S. Army in Vietnam, 1965–70*. Washington, D.C.: Department of the Army, 1973.

DISEASE: MIGRATION, CONQUEST, AND AFRICAN EXPLORATION

Ackerknecht, Erwin H. *History and Geography of the Most Important Diseases*. New York: Hufner Publishing Co., 1965.

Baikie, William Balfour. *Narrative of an Exploring Voyage Up the Rivers Kwo'ra and Bi'nue (Commonly Known as the Niger and Tsa'dda) in 1854*. London: John Murray, 1856.

Cameron, Ian. *The Impossible Dream: The Building of the Panama Canal*. London: Hodder and Stoughton, 1971.

Carlson, Dennis G. *Africa Fever: A Study of British Science, Technology and Politics in West Africa, 1784–1864*. Canton, Mass.: Science History Publications USA, 1984.

Cartwright, Frederick F., and Michael Biddiss. *Disease and History*. Guildford, Surrey: Sutton, 2000.

Childs, St. Julien Ravenel. *Malaria and Colonization in the Carolina Low Country, 1526–1696*. Baltimore: Johns Hopkins University Press, 1940.

Curtin, Philip D. *Disease and Empire: The Health of European Troops in the Conquest of Africa*. Cambridge: Cambridge University Press, 1998.

Diamond, Jared. *Guns, Germs, and Steel*. London: Vintage, 1998.

Gelfand, Michael. *Livingstone the Doctor, His Life and Travels*. Oxford: Basil Blackwell, 1957.

Hemming, John. *The Conquest of the Incas*. London: Macmillan, 1970.

Klein, Herbert S. *African Slavery in Latin America and the Caribbean*. New York and London: Oxford University Press, 1986.

McCullough, David. *The Path Between the Seas: The Creation of the Panama Canal (1870–1914)*. New York: Simon and Schuster, 1977.

Ransford, Oliver. *David Livingstone: The Dark Interior*. London: John Murray, 1978.

Acknowledgments

RESEARCHING THIS BOOK gave me a license to travel widely in South America and spend many contented hours browsing in libraries and archives I would never otherwise have entered. But without the advice and support of a number of people, I would not have gotten very far and the browsing would have lacked direction.

First, the logistics. For helping me navigate Venezuelan bureaucracy, the Orinoco, and the Casiquiare canal, my thanks to Alejandro Reig, Hugo Perdomo, and Sixto, and for the loan of her jungle-strength mosquito net, to Yasmin Rubio Palis. For guiding me over Chimborazo and nursing me through the dreaded *soroche*, I am indebted to the best expedition planners in Ecuador, Robert and Daisy Kunstaetter, and for their hospitality in Vilcabamba, to Orlando and Alicia Falco of the Ruimi Huilco nature reserve.

For chauffeuring me safely along some of the worst roads in Bolivia, and for their unfailing good humor, my thanks to Isaac and Eulogio and their Toyota That Could. In Colombia, I am indebted to Alberto Gomez Mejia and Jesús Idrobo of the Bogotá Botanical Garden for showing me the finest cinchona tree "in captivity," and to Dr. Manuel Patarroyo for taking time out of his busy schedule to fly with me to Leticia.

Now the facts. The research for this book was carried out principally at the Wellcome Trust Library in the Euston Road, the Royal Geographical Society in Kensington, and the Royal Botanic Gardens in Kew. For granting me access to Spruce's original notes and journals, I am particularly grateful to Kew archivists Kate Pickard and Sylvia Fitzgerald, and for access to Markham's journals, to Hugh Thomas, the archivist at the RGS. Thanks also to Lorraine Jones, assistant curator of the Royal Pharmaceutical Society of Great Britain, and Diana Wyndham, in New South Wales, for their help with sources on Ledger, and to Soni Veliz in London for Spanish translations.

For answering my many questions about the science of malaria and quinine, I am indebted to Professor David Warhurst of the London School of Hygiene and Tropical Medicine and to his colleagues Chris Curtis and Eleanor Riley. Similarly, my thanks to Robert T. Joy of the Uniformed Services University of the Health Sciences in Bethesda, Maryland, for guiding me through the voluminous military bibliography on malaria. All of the above also read early drafts of the manuscript for errors.

In addition, I am grateful for the comments of Dr. Fernando Ortiz-Crespo of FUNDACYT in Quito, and Jordan Goodman of the University of Manchester Institute of Technology. Although not every suggestion of theirs was adopted, together they saved me from making many embarrassing mistakes (any that remain are entirely my own).

Finally, this book would not have been possible without the expertise of my agent, Derek Johns, the skill of my editor, Jeremy Trevathan, and most of all the unfailing belief and support of my wife, Jeanette, and my parents, Frank and Naomi.

Index